Pursuing Scientific Humanism

Pursuing Scientific Humanism
Letters Between Werner Heisenberg and Enrico Cantore, 1967–1976

EDITED AND TRANSLATED BY
Claudio Tagliapietra

WITH A FOREWORD BY
Peter Harrison

CASCADE *Books* · Eugene, Oregon

PURSUING SCIENTIFIC HUMANISM
Letters Between Werner Heisenberg and Enrico Cantore, 1967–1976

Copyright © 2025 Claudio Tagliapietra. All rights reserved. Except for brief quotations in critical publications or reviews, no part of this book may be reproduced in any manner without prior written permission from the publisher. Write: Permissions, Wipf and Stock Publishers, 199 W. 8th Ave., Suite 3, Eugene, OR 97401.

Cascade Books
An Imprint of Wipf and Stock Publishers
199 W. 8th Ave., Suite 3
Eugene, OR 97401

www.wipfandstock.com

PAPERBACK ISBN: 979-8-3852-4329-7
HARDCOVER ISBN: 979-8-3852-4330-3
EBOOK ISBN: 979-8-3852-4331-0

Cataloguing-in-Publication data:

Names: Tagliapietra, Claudio, editor and translator. | Harrison, Peter, foreword.

Title: Pursuing scientific humanism : letters between Werner Heisenberg and Enrico Cantore, 1967–1976 / edited and translated by Claudio Tagliapietra ; foreword by Peter Harrison.

Description: Eugene, OR: Cascade Books, 2025 | Includes bibliographical references.

Identifiers: ISBN 979-8-3852-4329-7 (paperback) | ISBN 979-8-3852-4330-3 (hardcover) | ISBN 979-8-3852-4331-0 (ebook)

Subjects: LCSH: Heisenberg, Werner, 1901–1976. | Physics—Biography. | Cantore, Enrico. | Science—Philosophy. | Humanism. | Religion and science.

Classification: QC16.H35 T35 2025 (paperback) | QC16.H35 (ebook)

VERSION NUMBER 03/18/25

Archival Permissions:

Permission to publish the letters from the Heisenberg estate was granted by the Archives of the Max Planck Society, Berlin, which holds the complete rights of use for Werner Heisenberg's letters. The letters authored by Enrico Cantore and other correspondents may still be subject to copyright by their legal successors.

The publisher and author have made every effort to trace and obtain permissions for all copyrighted material included in this work. Any claim to material requiring further permissions should be addressed to the publisher for prompt resolution.

The author expresses gratitude to the Archives of the Max Planck Society for their invaluable assistance and for authorizing the inclusion of these historical documents.

I would hope that your new institute gives an opportunity to the younger generation to participate in fostering the use of science for building a more human world.

> Werner Heisenberg to Enrico Cantore
> February 20, 1974

Contents

Foreword by Peter Harrison | ix
Acknowledgments | xiii
Editorial Note | xv

Part I. Contextualizing the Correspondence

1. Werner Heisenberg and Enrico Cantore on Scientific Humanism: An Introduction to a Conversation in Letters and Meetings | 3
2. Enrico Cantore: An Intellectual Profile
   —by Giuseppe Tanzella-Nitti | 26
3. Enrico Cantore: Biography, Chronology, and Publications | 40

Part II. Correspondence (1967–1976)

I. "I have taken the courage to write to you." First Encounters, May 1967–October 1967 | 49

II. "New York desperately needs scientific humanism." The First Fordham Period and Heisenberg's Proposed Honorary Doctorate, December 1967–May 1968 | 77

III. "The philosopher needs the benevolent help of the creative scientist." Second Meeting and the Writing of Heisenberg's "Physics and Beyond," July 1968–June 1969 | 119

IV. "You tell me not to lose courage." Continued Support and Publication of "Physics and Beyond," June 1969–May 1970 | 169

V. "I feel profoundly close to you in spirit": Enrico Cantore Gaining Traction, December 1970–December 1972 | 197

VI. "I wish to convey to you some encouraging news." Final Letters and Positive Steps Forward for Scientific Humanism, August 1973–June 1976 | 239

Epilogue | 267

Appendix. The "Cantore Affair" at Fordham University | 271
Glossary of Names | 283
Bibliography | 291

Foreword

IN HIS 1959 REDE Lecture, British scientist and novelist C. P. Snow introduced the now familiar idea of the "two cultures," lamenting the growing divergence in Western society of the natural sciences and the humanities. The correspondents whose absorbing exchanges form the subject matter of the present book, each in his own way, advocated for an overcoming of this divide. Theoretical physicist Werner Heisenberg will be well known to readers. A pioneer of quantum mechanics, for which he was awarded the Nobel Prize in 1932, he established the eponymous "uncertainty principle," which encapsulates one of the deep mysteries of quantum mechanics: we can never simultaneously know both the momentum and position of a sub-atomic particle. This principle confounds common sense and has far-reaching implications for the philosophy of science. Heisenberg himself concluded, on the basis of his work in quantum mechanics, that we can never have direct access to sub-atomic reality, and that accordingly we need to relinquish the ambition to an objective knowledge of the fundamental entities that make up our world. Quantum physics, then, inevitably leads to philosophical questions about the limits of science and of human knowledge more generally.

Our other correspondent, the Jesuit philosopher Enrico Cantore, will be less familiar to readers. Until encountering this rich correspondence I knew virtually nothing about Cantore and am grateful to Claudio Tagliapietra for his labors in compiling, cataloguing, and translating this correspondence. In doing so he has performed a sterling service, introducing to us a relatively little-known figure whose ideas and aspirations are deserving of wider circulation. The science and religion field, in its own way a response to the "two cultures" problem, tends to operate with a well-established, and somewhat overworked *dramatis personae*, especially in the Anglophone world. It is refreshing to have added to the usual suspects a relatively novel character, and to be able to reflect on what new insights his thought might bring to the field.

Cantore used the label "scientific humanism" to characterize his general approach. "Humanism" in this context needs some unpacking, since in some of its familiar manifestations—in the forms of "secular humanism" or "exclusive humanism," for example—it represents a trend that is hostile or, at best, indifferent to religion. Humanism in this latter register seeks to establish a way of being in the world that makes no reference to realities beyond the mundane and material. Such humanism is often prone to scientism. By way of contrast, the scientific humanism espoused by Cantore sought to expand the horizons of the sciences, in order to arrive at forms of understanding that incorporated the interests of the humanities and to some extent the wisdom traditions of the world. The "humanism" in question is thus the humanities, broadly conceived. It was Cantore's ambition, to use his own words, to "bridge the gap between modern science and the old humanistic philosophical tradition of Europe." Integral to this vision was the ideal of a holistic education that married the virtues of each of the "two cultures."

As the following correspondence makes abundantly clear, Heisenberg was deeply sympathetic to Cantore's aspirations. While some physicists of more recent vintage (Richard Feynman, Lawrence Krauss, Stephen Hawking, Neil deGrasse Tyson) have expressed an indifference to, or even contempt for, philosophy, Heisenberg was fully aware of both the philosophical foundations of physics, and of the broader philosophical implications of quantum mechanics. This was partly the result of an intellectual formation in the classics, including a reading knowledge of ancient languages. Heisenberg maintained a lifetime interest in Plato, and it is likely that his understanding of fundamental physics was informed by Kantian philosophy.[1] As his paper "Scientific Truth and Religious Truth" indicates, he was also possessed of a sensitivity to religious questions, which perhaps accounts for his willingness to engage with Cantore and support his efforts. Adverting to the notorious trial of Galileo and the shadow it cast on the subsequent history of science–religion relations, he once wrote: "I have never been able to dismiss the content of religious thinking simply as a stage in human consciousness which we have superseded, as a part

---

1. See Camilleri, "Heisenberg and the Transformation of Kantian Philosophy."

which we can dispense with in future. So I have continually been forced during my life to ponder on the relationship between these two worlds of the spirit [science and religion], for I have never been able to doubt the truth of what they are pointing to."[2]

While, surprisingly perhaps, the correspondence does not dwell much upon spiritual matters—neither correspondent seems to have raised specific religious questions—there is a clear convergence of views about the impoverished state of philosophy in the English-speaking world, and the undesirable dominance of the "empiricist positivist" tradition. This was seen to contribute to a widespread view of science as "inhuman and dehumanising," to use Cantore's characterization. Both figures were deeply puzzled by the success of reductionist works of popular science—such as molecular biologist Jacques Monod's *Chance and Necessity* (1970) and behaviorist psychologist B. F. Skinner's *Beyond Freedom and Dignity* (1971)—which exemplified the sclerotic philosophical approach they both deplored. Science, they agreed, needed to be re-humanized.

While Heisenberg emerges in these exchanges as very much the senior figure, the correspondence worked to their mutual advantage. On occasion, Cantore was able to correct some of Heisenberg's misconceptions about the history of science—for example, his uncritical acceptance of the appealing but mistaken account of Galileo dropping weights from the Tower of Pisa. On the other side, when Cantore needed to test whether his depiction of scientific creativity was true to the experience of working scientists, he was able to confirm this with one of the leading scientific figures of the age. Cantore also sponsored Heisenberg for an honorary doctorate from Fordham (which for various reasons fell through). For the most part, though, the correspondence bears testimony to Heisenberg's unfailing generosity in supporting Cantore's attempts to gain a firm foothold in US academe (which met with varying degrees of success).

In addition to exposure to the ideas of these two individuals, the correspondence also provides insights in the vicissitudes of the academic life, and Cantore's in particular. Cantore struggled for recognition and was often disappointed that his published work seemed to have had little impact on the scientific community. His quest to establish an institute at Fordham met with resistance or indifference—perhaps a manifestation of the very problem of "the two cultures" that he sought to address. Yet he had some successes: the publication of his

---

2. Heisenberg, "Scientific Truth and Religious Truth," 463.

books and papers, and the eventual establishment of the Institute for Scientific Humanism in New York in 1973, the spirit of which continues in the International School for Interdisciplinary Research at the Pontifical University of the Holy Cross, Rome. This present publication of his correspondence with Heisenberg is another measure of his ongoing influence. Not only does it provide rewarding insights into the lives, struggles, and ideals of two remarkable individuals: it also reminds us of the enduring importance of Cantore's mission.

Peter Harrison
Brisbane
February 2025

Acknowledgments

My most heartfelt thanks go to the Archiv der Max-Planck-Gesellschaft zur Förderung der Wissenschaften in Berlin for its permission to examine the correspondence of Werner Heisenberg and to use this exceptional archive material. I am particularly grateful to Florian Spillert for his professional support during the research process. I am also much indebted to Gabriella DiMeglio, Archives and Special Collections Librarian at Fordham University, for her practical help during my research at the Fordham University Archive. My work also greatly benefited from the support of the Librarians and Staff of the Library of Congress, Washington, DC, where I first discovered the existence of the correspondence held at the Max Planck Institute in Berlin. I also thank the Case Western Reserve University Archives for the valuable information and the University of Turin and Pontifical Gregorian University archives for providing academic information on Enrico Cantore. I owe gratitude to Paul Vitz for his great support with rich personal testimony about Enrico Cantore and the academic environment in New York during the seventies.

I want to give special thanks to Giuseppe Tanzella-Nitti of the Pontifical University of the Holy Cross of Rome, who made possible my access to the Papers and Manuscripts of Enrico Cantore, housed at the Center for Interdisciplinary Documentation on Science and Faith (DISF Center). I also thank Brandon Vaidyanathan for hosting me at the Department of Sociology of The Catholic University of America during the summers of 2023 and 2024. This arrangement offered the perfect environment for work on this manuscript.

I am also indebted to the residents of Tenley Study Center for becoming my family in Washington, DC, and providing the milieu of work and prayer, without which this project could not have been realized.

I am deeply grateful to Professor Peter Harrison for his generosity in contributing a foreword, which enriches the volume with valuable historical and philosophical context.

Finally, very special thanks are also owed to Leigh Biddlecome, whose creative and wise insights, professional help, and editorial assistance made it possible for this project to see the light of day.

Editorial Note

THIS MANUSCRIPT IS THE result of research carried out on Werner Heisenberg's personal correspondence, hosted at the Archive of the Max Planck Society in Berlin. It is a collection of 107 letters, written between May 2, 1967, and June 29, 1976, which I uncovered in the archives of the Max Planck Society. Not all of the letters were exchanged between Heisenberg and Cantore; I have also included letters from other individuals who played a more or less significant role in the narrative, such as Wolfgang Büchel, Wolfgang Wickler, Patrick A. Heelan, Ruth Nanda Anshen, Fordham University, and Annemarie Giese.[1]

The material is organized into six chapters, each forming a distinct narrative unit. A brief editorial overview introduces each unit, and I have included footnotes to help readers engage with the letters more fully. The book concludes with an Epilogue.

Another important set of documents comes from the Fordham University Archives, where I conducted research to document a different perspective on the existence of the Institute for Scientific Humanism, the failed attempt to confer an honorary doctorate on Heisenberg, and Cantore's dismissal. These documents form the basis of the appendix about the "Cantore Affair" at Fordham University, which comes after the Epilogue, as it does not directly relate to Cantore's correspondence and represents an independent perspective on the events.

---

1. Of the 107 letters, 67 were exchanged between Heisenberg and Cantore (45 sent by Cantore and 22 by Heisenberg), while the remaining 40 were exchanged between Cantore, Heisenberg, and other individuals.

In addition, I have also included an intellectual biographical essay on Enrico Cantore by Giuseppe Tanzella-Nitti, a close friend and successor to Cantore's vision of scientific humanism.

The translation of these letters has not been easy. This is partly because some were originally written in English by non-native speakers (Cantore and Heisenberg), partly because others were translated into English from German, and partly because one of the correspondents (Cantore) wrote in German, which was not his native language.[2] I have taken great care to maintain the authenticity and integrity of their original communications.

My deliberate aim in translating Heisenberg's German letters to American English was to save as much of the original tone, nuance, and choice of terms as possible. Indeed, Heisenberg's language is very precise and reflective, also due to the mediation of his secretary, Ms. Giese. I tried to make sure that the same intellectual rigor and the subtlety of his thoughts were preserved. This meant that I had to keep as close as possible to the literal meanings of his words while trying to echo the cadence and formality that characterized his writing style. A similar comment applies to the ten English letters signed by Heisenberg.

On the other hand, Enrico Cantore wrote his letters in German and English, and his way of expressing himself bears the unmistakable imprint of his Italian mother tongue. Where I have "Englished" the German letters using the same rule as in Heisenberg's case, I did so without "correcting" Cantore's English, even where standardization would have made for greater clarity. This decision was made to recognize the unique voice that emerges from his linguistic choices. The occasional idiosyncrasies and deviations from conventional English usage are not merely linguistic quirks; they reflect Cantore's nature and the feelings that pervade his letters. In their raw, unfiltered form, his words give readers a direct connection with the man behind the words.

I hope that the readers of these letters will get as close as possible to their original form and gain an impression of the intellectual exchange and the deep respect Heisenberg and Cantore had for each other. The letters are more than historical documents: they are personal reflections from two great minds in conversation, and it is my privilege to bring their voices to the page in a way that honors their original intentions.

Claudio Tagliapietra
December 2024

---

2. Of the 107 letters, 47 are written in German: 14 by Cantore, 33 by natives, including Heisenberg.

PART I

Contextualizing the Correspondence

1

Werner Heisenberg and Enrico Cantore on Scientific Humanism

An Introduction to a Conversation in Letters and Meetings

CLAUDIO TAGLIAPIETRA

BERLIN, SUMMER 1967. WERNER Heisenberg finds a letter on his desk. Its opening line—"Dear Professor Heisenberg"—might have seemed ordinary. The sender was Enrico Cantore, an Italian Jesuit who had recently moved to the US to teach philosophy of science at Fordham University. What could a Jesuit philosopher and the founder of quantum mechanics possibly have in common? Why did Enrico Cantore decide to write specifically to Heisenberg? And why did Heisenberg continue responding to his letters? What can we learn from these letters today? These and many other questions filled my mind when, to my great surprise, I stumbled upon this correspondence almost by chance.

Werner Heisenberg with students, Student evening during the Ninth Conference of Physicists at Lindau City Hall, 1959.[1]

As I sifted through the letters, I not only uncovered the facts behind publications like *Physics and Philosophy* (1958) and *Physics and Beyond* (1971) by the great German physicist, but I also came to know intriguing details of Cantore's academic life and his major works, such as *Atomic Order* (1969) and *Scientific Man* (1977). Gradually, an intellectual exchange took shape—a story of support and encouragement that the renowned German physicist chose to give to an idealistic researcher, nurturing his philosophical passion for interdisciplinary dialogue between scientists and philosophers. Moreover, Heisenberg fostered a very concrete initiative to make this dialogue possible, starting from the university environment—the only setting that, traditionally, has preserved the social mission of cultivating such dialogue.

Cantore was so convinced of the philosophical stature of Heisenberg's scientific thought that he went to considerable effort to have the physicist awarded an honorary doctorate in Humane Letters from Fordham. Following a precipitous turn of events that led to Cantore's dismissal from the prestigious university, Heisenberg declined the honor. As their

---

1. License CC BY. State Archives of Baden-Württemberg, Germany (W 134 Nr. 058165d). Photo: Willy Pragher.

mutual acquaintance progressed, Heisenberg repeatedly demonstrated his support both for the humanistic-scientific ideal of the Jesuit philosopher and for the man himself. He tried in many ways to help Cantore secure an academic position and promote his publications. Heisenberg's trust for him went so far as to propose Cantore's name as translator from German into English for his autobiography *Physics and Beyond* at the famous publishing house Harper & Row.

Without Heisenberg's support, Cantore would not have been able to realize the dream to which he had dedicated his life. Anyone who works in academia can likely attest to this, recalling how the generous support and feedback of a senior scholar, even when faced with challenges or opposition from colleagues, has positively influenced their intellectual growth and academic output. This may be one of the most important lessons from this story—for senior scholars and young academics at the beginning of their careers, who can learn much from this account and find the courage to invest their lives in pursuing great ideas.

The story begins with a tentative exchange of manuscripts, develops alongside Cantore's career, and crosses his academic, financial, practical, and theoretical challenges, ultimately leading to an unexpected conclusion. Cantore's life teaches us about the persistence and tenacity required to pursue great ideas. This demands from the humanist scientist the qualities of courage and fidelity to their ideals, even in the face of unexpected or unforeseeable challenges and difficulties.

The ideas behind this correspondence could enrich the understanding of anyone in academia or beyond who conducts research today, particularly those seeking to highlight the humanistic aspects of science, regardless of the specific field of inquiry: science is a human endeavor to promote human dignity. There is a significant risk of practicing an inhumane, dehumanizing, or even outright dehumanized science, especially in an age where artificial intelligence is increasingly encroaching upon fields already governed by digitalization and automation.

Scientific Humanism: The Urgency of a Bridge Between Scientific and Humanistic Culture

What could a Jesuit philosopher and the founder of modern quantum mechanics possibly have in common? Behind this correspondence lies a powerful idea—scientific humanism. As Heisenberg wrote in the introduction to his autobiography, *Physics and Beyond*, "Science is made by man . . . science is rooted in conversation . . . science is quite inseparable from these [human, philosophical, or political] broader questions." Science is not merely an expression of humanity's investigative power over nature, where discoveries are used to dominate both nature and others through exploitative omnipotence. It is a human endeavor, whose ultimate purpose is to *humanize*—to enable humanity to reach its full potential. This fullness, in a certain sense, is something scientists owe to humanity. Through their work, scientists are responsible for creating conditions that allow every individual to achieve what is rightfully theirs: their own dignity. In this way, the discoveries scientists make in their exploration of reality form the essence of their mission, their "vocation."

According to Cantore, this task must take the form of a "humanizing dialogue" between science and philosophy, a necessary dialogue in an age of heavy technologization, mechanization, and automation, where philosophical reflection on the consequences of scientific work for humanity is often sidelined. This division tends to separate specialized scientific knowledge from humanistic culture, promoting the belief that the two realms yield distinct practices, lifestyles, and mentalities that are incompatible and mutually unintelligible: empirical knowledge, regarded as objective and indisputable, versus humanistic knowledge, which emphasizes reflective and interpretive understanding, considered subjective. This is the "divide between the two cultures" theorized by C. P. Snow in the late 1950s, which was deemed to lead to the mutual isolation of philosophers and scientists. Both Cantore and Heisenberg viewed this divide as one of the tragedies of modernity and one of the greatest dangers to the future of civilization.[2]

By the late 1960s, little thought had been given to how this rupture might be healed. Interdisciplinarity was a concept only beginning to resurface, having been largely forgotten by a university tradition that favored hyperspecialization over diversification of academic curricula. The philosophy of science was still a relatively new field in universities, and it was unclear whether it should be practiced by scientists in the

---

2. "The issue of the significance of science for man is certainly one of the most critical of our civilization. The embarrassing phenomenon of the so-called two cultures stems from it. Many people nowadays object to science because they consider it inhuman and dehumanizing. Many others, including numerous scientists, feel bewildered and discouraged. I share the conviction that science—if properly understood and followed through—does indeed constitute a humanizing process for man." Cantore, "Science as Dialogical Humanizing Process," 293.

role of philosophers or philosophers trying to understand scientists. Meanwhile, society was feeling a profound need for humanity, less than twenty years after the detonation of the first nuclear bomb on a civilian population and the horrors of a war that had demonstrated humanity's capacity to destroy itself.

Modern physics was in its golden age. One need only look at the photograph of the Fifth Solvay Conference in 1927, where seventeen of twenty-nine individuals had won at least one Nobel Prize.[3] The progress of humanity seemed, at once, both bright and promising yet also dark and terrifying.

Fifth Solvay Conference, 1927. First row: Langmuir, Planck, Curie, Lorentz, Einstein, Langevin, Guye, Wilson, Richardson. Second row: Debye, Knudsen, W. L. Bragg, Kramers, Dirac, Compton, de Broglie, Born, Bohr. Third row: Piccard, Henriot, Ehrenfest, Herzen, de Donder, Schrödinger, Verschaffelt, Pauli, Heisenberg, Fowler, Brillouin.[4]

With this context in mind, we can better understand the theoretical challenges Enrico Cantore faced in fulfilling the mission of scientific humanism, as reflected in his writings today. We may also better grasp the idealistic fervor that permeates his work, which echoes the calls and

---

3. Since 1911, the Solvay Conferences (French: Congrès Solvay) have been dedicated to addressing the most important unsolved problems in both physics and chemistry. Marie Curie was the only person to have won the Nobel Prize in two different disciplines: physics and chemistry.

4. This image is used under the Creative Commons license.

aspirations of the Catholic Church in urging the technological world not to forget the human roots and humanizing goal of human activity in the universe, as expressed through the Second Vatican Council (1962–1965). Cantore anticipated many of the insights of this Council, particularly those passages in the pastoral constitution *Gaudium et Spes* (1965) that address scientific and technological progress.[5] The Council affirmed that scientific progress should aim at the integral development of the human person and that we must deeply examine the meaning of culture and science for humanity.[6]

Another important conciliar influence that underpins Cantore's intellectual work is the conviction that scientific work must promote the dignity of the human person.[7] Just as the sage in ancient biblical culture had been a catalyst for such dialogue, the modern scientist must now assume that role—an innovator and staunch defender of human dignity. Cantore saw the figure of biblical wisdom and the sage as the archetype of the modern scientist. In this continuity, he recognized an ancient mission of divine origin, a responsibility, a "leadership," and a true "vocation."[8]

This is the theoretical foundation on which Cantore's Christian theological reflection takes root and where the Jesuit philosopher finds the ultimate impetus for his mission in service to the Catholic Church. In 1969, he wrote, "I think it a duty of mine toward the scientific world, and also a contribution to help the Church be brought up to date in a very significant area."[9]

We can only imagine the enthusiasm and spiritual drive he must have felt, knowing he was supported by the conclusions of the Catholic Church's teachings in the mission he had seen as his own from an early age. Since his youth, he had been convinced that he was called to promote scientific humanism and dedicate his life to bridging the gap between the two cultures.[10] However, this gap needed to be addressed

---

5. See Vatican Council II, *Gaudium et Spes* nos. 33, 36, 44.

6. See Vatican Council II, *Gaudium et Spes* no. 61.

7. See Vatican Council II, *Gaudium et Spes* nos. 12–22, particularly the dialogue with scientific atheism (nos. 19–21) and the Christian dignity of human labor (no. 22).

8. See Cantore, "La sapienza biblica, I–III."

9. Letter 48 (February 20, 1969). That scientific work has, as both its driving force and its ultimate goal, the promotion of human dignity is far from an abstract or outdated concept. Even today, the meaning of human dignity and what constitutes "human dignity" remains a topic of discussion, particularly in the field of bioethics.

10. From the time of his religious formation at nineteen years old (1945), Cantore had developed a deep spiritual conviction that he had received a mission: "The central

starting within the academic world, given the university's specific mission towards science and society.[11]

This is why, in 1967, Cantore moved to the United States to join Fordham University. Initially, the plan, agreed upon with the superiors of his religious order, was to further his studies in the philosophy of science and to integrate into academia. While Cantore saw promise in this path, he found it too narrow. As he observed, American philosophy was strongly influenced by the analytic tradition, and pursuing one of the traditional branches of philosophy, even continental philosophy, would have jeopardized the project of scientific humanism. This initiative, which began under the auspices of the Teilhard Institute already established at Fordham, would require the creation of a new institute. This institute would need fresh ideas, the active participation of faculty from various departments, particularly the sciences, long-term funding, and a "catalyst"—someone fully committed, body and soul, to fostering dialogue between scientists and philosophers. The goal of this endeavor was to enrich philosophy with the contributions of science and to make the scientific world more human. In short, the mission was "helping man find self-understanding in an age of science."[12]

Why Did Cantore Write to Heisenberg?

In the preface to his work on the humanistic significance of science, Cantore expresses his enduring gratitude to Heisenberg, calling him a true "scientist-humanist." This dedication reveals how, in Werner Heisenberg, Cantore saw the embodiment of the ideal scientist-humanist—a man of science who had managed to unite the two cultures into a marvelous whole, contributing equally to physics and philosophy.

At first, Cantore was not merely seeking out the scientist or scholar, the scientific leader he would eventually address in later writings. Initially, he sought contact with the "creative scientist."[13] This was a typical expression of Cantore, used to describe a scientist capable of exercising

---

goal of my life is to contribute to the solution of the so-called problem of the two cultures," he would write in an unpublished memoir in 1972.

11. See "Memorandum on Scientific Humanism," Letter 21 (February 17, 1968), *Attachment*.

12. Letter to MIT Press's first director, Carroll Bowen (June 11, 1969), attached to Letter 67 (June 27, 1969).

13. Cantore, *Scientific Man*, xvii–xviii.

"scientific creativity," an intellectual courage that leads to the formulation of groundbreaking theories and their validation through discovery, often resulting in revolutionary advancements. The examples Cantore mentions in his writings include Columbus, Galileo, and now Heisenberg. As Cantore described it, discovery is a process that requires courage, tenacity, and endurance to suffer for an ideal that demands total dedication and sacrifice. According to Cantore, the virtues required of the creative scientist are: attraction to an ideal, self-dedication to research, encounter with the Absolute, personal "coresponsibility," and "responsive communion."[14] Creative scientists need to cultivate such virtues not only for the sake of their immediate field of research but also of society at large. And Heisenberg embodied all of these.

For this reason, it was important for Cantore to engage in dialogue with Heisenberg, learn from him, and present him as an example. He wanted to tell the story of science not only through textbooks and articles but also through Heisenberg's life itself. Cantore sincerely appreciated Heisenberg's autobiographical approach in *Physics and Beyond*, mainly how it used the personal experiences of scientists at work as the source of scientific discovery. Cantore shared this approach in some of his own writings, which employed the "genetic method" he had learned from reading Jean Piaget.[15] This method, which Cantore would later pass on to his collaborators after founding the Institute for Scientific Humanism, was adopted by figures like Friedrich Trinklein, who in 1971 published a collection of interviews with scientists on the relationship between science and religion titled *The God of Science* (1971).[16]

When Heisenberg received Cantore's first letter, he was sixty-five years old and a global authority in quantum mechanics, a field he had helped establish and which had earned him the Nobel Prize in Physics

---

14. "Corresponsibility" and "responsive communion" are neologisms coined by Enrico Cantore. See Cantore, "Science as Dialogical Humanizing Process."

15. Cantore writes: "Drawing on his experience as a key participant in the creation of modern quantum theory, Heisenberg intends in this work to convey to the general public his profound conviction about the humanistic significance of science. The originality of the book consists in the fact that he makes his point not through theoretical discussions but by means of concrete testimonies expressed in dialogue between himself and the leading scientists with whom he worked." Cantore, "Science as Dialogical Humanizing Process," 293. Regarding the use of the genetic method, see Cantore, "La scienza come fattore umanistico."

16. See Trinklein, *God of Science*.

in 1932. He had been director of the Max Planck Institute for Physics in Munich since 1942.

Between 1939 and 1942, the institute had been involved in the German nuclear program. Despite the many offers he received during the war, Heisenberg chose to remain in Germany. His inevitable involvement made more than a few question his political affiliations and moral choices during the war years. Although evidence suggests that Heisenberg was not a supporter of Hitler's regime and may have deliberately slowed the progress of the nuclear program, his decision continued to cast a shadow on his reputation, particularly in postwar America during the tense years of the Cold War. In his review of *Physics and Beyond*, Cantore does not hesitate to defend his mentor's choice. He even sees in it a great scientific virtue: the personal corresponsibility of the scientist. Cantore writes, "Heisenberg testifies that his entire decision of not leaving Germany can be traced back to a single, short sentence of Planck's: 'Think of the time after the catastrophe.'"[17]

The institute, founded in Berlin and headquartered in Göttingen, had moved in 1955 to Garching, near Munich, where Heisenberg established his academic and personal home and where he would die on February 1, 1976, from kidney cancer that had now spread to his gallbladder.[18] Heisenberg stepped down as director of the Institute on December 31, 1970, and later relinquished his presidency of the Alexander von Humboldt Foundation in October 1975. These events are only hinted at in the correspondence we are now examining.

The monumental success Heisenberg achieved in physics did not absorb him to the point where he neglected the philosophical implications of his work. On the contrary, his classical education, artistic and musical interests, and character naturally inclined him toward high-level philosophical engagement, deeply influencing his scientific approach.

His well-documented biography reveals that he received a traditional German education, which included rigorous training in classical languages (his father was a professor of Greek at the University of Munich) and the study of classical literature. Early exposure to the works of ancient philosophers, particularly Plato, instilled in the young Heisenberg a profound appreciation for philosophical inquiry and the

17. Heisenberg, *Physics and Beyond*, 154, cited in Cantore, "Science as Dialogical Humanizing Process," 310–12.

18. By a remarkable coincidence, Enrico Cantore would pass away thirty-eight years later from the same illness as Heisenberg.

intellectual traditions of the West.[19] This interest was further cultivated during his university studies in physics under Arnold Sommerfeld's guidance. Heisenberg's fascination with Plato's idea of a structured universe, as described in the *Timaeus*, later influenced his scientific work, particularly in his search for the fundamental principles of quantum mechanics. His classical training and philosophical interests, combined with his understanding of the physical world, enabled him to reflect deeply on the implications of his scientific discoveries, especially the uncertainty principle (1927), and to communicate these ideas in a way that was accessible to both scientists and philosophers.

Heisenberg maintained an active interest in philosophical discussions throughout his career, often engaging with contemporary philosophers and theologians. His dialogues with figures such as Martin Heidegger and Carl Friedrich von Weizsäcker, both of whom were close friends, further demonstrate his ongoing commitment to exploring the philosophical dimensions of science. Heisenberg had a long-standing friendship with Heidegger and was familiar with the German philosopher's hermeneutic phenomenology and his critique of Greek philosophy, contributing an essay, "Fundamental Prerequisites in the Physics of Elementary Particles," to the *Festschrift* in honor of Heidegger in 1959.[20]

Heisenberg's passion for philosophy and his inclination to reflect on the philosophical implications of science led to a parallel stream of philosophical publications alongside his scientific work. Among his most important contributions in this field, available in English translation, are the eight lectures delivered between 1932 and 1948 that make up *Philosophic Problems of Nuclear Science* (1952);[21] the Gifford Lectures given at the University of St. Andrews in the winter of 1955–1956, published as *Physics and Philosophy: The Revolution in Modern Science* (1958); his autobiographical interviews in *Physics and Beyond: Encounters and Conversations* (1971);[22] and, with a greater focus on the relationship between scientific and religious language, the essay "Scientific Truth and Religious Truth" (1979), based on a lecture Heisenberg gave

---

19. See Cantore, "Science as Dialogical Humanizing Process," 297–98, which cites Heisenberg, *Physics and Beyond*, 8–14.

20. See Heisenberg, "Grundlegende."

21. See Heisenberg, *Philosophic Problems*.

22. See Heisenberg, *Teil*.

in Munich in 1973 when he received the Romano Guardini Prize from the Catholic Academy of Bavaria.[23]

There is a striking continuity in Heisenberg's "interdisciplinary method" across his philosophical writings. A glance at the table of contents in *Philosophic Problems of Nuclear Science*, his earliest philosophical work, reveals how he could traverse classical philosophy and nuclear physics with an impressive capacity for dialogue. For example, in one of the early letters, Cantore expressed his admiration for Heisenberg's essay on Goethe's theory of color, a theory influential in physiology, art, and aesthetics, which Heisenberg compared to Newton's corpuscular theory of light—criticized by Goethe.[24] The essay delved into Goethe's biographical details to explain the development of his theory of colors, demonstrating how scientific theories are always the product of personal intellectual endeavors, shaped by the scientist's experience. The result is that theories can explore reality through different, seemingly contrasting but ultimately complementary, pathways. Heisenberg began *Physics and Beyond* with this very idea: "Science is made by man."

As for Heisenberg's figure, this correspondence reveals new insights beyond what has been previously published. For instance, the letters show that Fordham University, at Cantore's suggestion, had decided to award Heisenberg an honorary doctorate in Humane Letters. However, following the "Cantore Affair" and the university's inability to help Cantore secure a position at the prestigious institution, Heisenberg declined the honor. Another previously unknown aspect is the editorial story surrounding *Physics and Beyond*. Heisenberg had proposed Cantore as the translator for the work from German into English, and the correspondence serves as a key to understanding the relationship between Heisenberg and the prestigious publishing house Harper & Row, particularly with Ruth Nanda Anshen, editor of the *World Perspectives* series, who became a friend and correspondent of Heisenberg.[25] Ultimately, after careful consideration of Heisenberg's suggestions and Cantore's abilities as a translator, the publisher decided to entrust the translation to Arnold

23. See Heisenberg, "Naturwissenschaftliche"; Letter 100 (August 30, 1973).

24. See Heisenberg, "Teachings of Goethe and Newton."

25. Heisenberg writes Anshen: "I would be happy if the translation could be entrusted to Father Cantore, as he deserves support and encouragement, even aside from our specific project. His knowledge of German is certainly sufficient, but I must leave the evaluation of his English skills entirely to you, as you have a much better and more reliable judgment in this regard." Letter 52 (March 25, 1969).

Pomerans. Heisenberg expressed his regret to Cantore, acknowledging that the decision ultimately rested with the publisher, and assured him that he would rely on him for the English translation of another important essay, "The Meaning of Beauty in the Exact Sciences," which Heisenberg had delivered at the Bavarian Academy of Fine Arts in 1971.[26]

Another new element is the portrayal of Heisenberg's more personal and human side, particularly how he interacted with collaborators and colleagues—an aspect that biographies often struggle to capture, given the richness of a life like that of one of the twentieth century's most influential minds. This dimension emerges most clearly during moments of emotional tension in Cantore's correspondence, especially regarding the events that unfolded at Fordham and within the New York academic scene, adding an element of interest to the letters.

Why Did Heisenberg Write to Cantore?

We might wonder why Heisenberg responded to Cantore's letters. After all, many philosophers and scientists had investigated Heisenberg's philosophical ideas and maintained active correspondence with him. Among these correspondents, as we'll see, were several Catholic priest-scientists. Cantore, therefore, was not the only philosopher-priest with whom Heisenberg corresponded.

---

26. Heisenberg, "Bedeutung."

Werner Heisenberg (center, with his wife on the right) was honored with the "Romano Guardino Prize" by Dr Franz Henrich (on the left), leader of the Catholic Academy of Bavaria, on March 25, 1973, Munich.[27]

One figure worth mentioning in this context is Patrick A. Heelan, a Jesuit and professor of philosophy, first at Fordham University, then at the State University of New York, and finally at Georgetown. Heelan is cited multiple times in the documents gathered in this volume. During his studies in Dublin in the late 1940s, Heelan had already worked with Erwin Schrödinger and John Synge, and in the early 1960s, he collaborated with Heisenberg while pursuing his philosophical studies at the Husserl Archives in Leuven, Belgium. Heelan maintained a lively correspondence with Heisenberg until the physicist's death in 1976. His philosophical work on the significance of quantum mechanics was deeply marked by his immersion in phenomenology, hermeneutics, and transcendental epistemology. His relationship with Heisenberg led to the writing of two major works: *Quantum Mechanics and Objectivity: A Study of the Physical Philosophy of Werner Heisenberg* (his doctoral dissertation, published in 1965) and *The Observable: Heisenberg's Philosophy of Quantum Mechanics*, completed in 1970 but published only posthumously in 2015.[28]

27. Istvan Bajzat/picture alliance via Getty Image.

28. Heelan, "Phenomenology, Ontology, and Quantum Physics." Regarding the reception of Heelan's book, Heisenberg remarked to Wolfgang Büchel: "It seems to me that the most important thing at the moment is that these problems are being seriously thought about at all; the fact that they are being thought about is more important than the particular result the thinker comes to." Letter 29 (April 2, 1968).

These examples demonstrate how Heisenberg's thought, along with his support and personal engagement, shaped the reflections of several philosophers, many of them, by chance, Jesuits—including, of course, Enrico Cantore. Their collaboration was clearly professional, with Heisenberg acting as a senior scholar and mentor. There is no evidence of any religious or spiritual references in their exchanges.

It is difficult to reconstruct from biographical sources and Heisenberg's sparse writings what he thought of religion, particularly Catholicism. In his biography of Heisenberg, David Cassidy notes that Werner's mother, Annie, converted from Roman Catholicism to her husband Augustus's Lutheran faith before their marriage.[29] The environment in which Heisenberg grew up was deeply religious, at least ethically, though with a certain ambivalence. His parents allowed their children to follow or reject religious norms, which Heisenberg later remarked were regarded more as empty appearances.[30] In an interview, Heisenberg once said: "My parents were far away from the Christian religion as far as the dogmas were concerned, but they would always stick to the Christian ethics. They would accept the rules of how to behave and to live and say that we can take them from the Christian religion, but we cannot accept literally all these old stories."[31]

Thus, Heisenberg was raised to view religion as adherence to external moral and social norms, not as faith in a personal God. However, Heisenberg's view of religion evolved, as reflected in *Physics and Beyond* (1971), which includes a dialogue from 1927 between him, Pauli, and Dirac on the validity of religion and the positions Einstein and Planck had expressed during the Fifth Solvay Conference. Pauli and Heisenberg believed that religion encompassed an area that involved the whole person and society, not just a subjective dimension.[32] Heisenberg expressed dissatisfaction with the separation of science and religion.[33] In his adult life and again toward the end of his life, Heisenberg declared that science and religion were "complementary" aspects of reality, each with its own

---

29. Cassidy, *Uncertainty*, 7–8.

30. Heisenberg, letter to his parents, January 11, 1928, cited in Cassidy, *Uncertainty*, 13.

31. "Double Dialogue with Werner Heisenberg," 475, cited in Cassidy, *Uncertainty*, 13.

32. Heisenberg, "Religion and Science (1927)."

33. Heisenberg, *Physics and Beyond*, 83, cited in Cantore, "Science as Dialogical Humanizing Process," 303.

language and symbolism, and each with its own limited sphere of validity. Religious and intuitive truths, he believed, should be seen as different facets of the same truth, while rational science—his own profession—was just one of many ways to perceive reality.[34] Cassidy reports that shortly before his death, Heisenberg told his longtime colleague and confidant Carl Friedrich von Weizsäcker: "If someone were to say that I had not been a Christian, he would be wrong. But if someone were to say that I had been a Christian, he would be saying too much."[35]

Despite his ambivalence toward religion, and within the context of relationships that were always maintained on a professional or intellectual level, it's worth noting that Heisenberg had numerous Catholic philosophers and scientists in his orbit, many of them Jesuit priests. Even the documents included in this text point to such connections. We do not have evidence that Cantore and Heisenberg ever discussed religion, but Cantore occasionally assured the scientist of his prayers for him, his wife, and his secretary in the closings of some letters. We can certainly say that Heisenberg had no objections to engaging in dialogue with Catholic scientists and philosophers who were seriously interested in understanding reality, and these individuals were similarly keen on learning his worldview. Starting with his friend Heidegger, who had received Jesuit training in his youth and studied Catholic theology for two years, Heisenberg also interacted with Patrick Heelan, Michael Yanase, and Wolfgang Büchel, to name just a few figures mentioned in the correspondence. For this reason, it's reasonable to assume that the letter from the young Jesuit physicist Cantore was not an unexpected novelty for the great physicist and was, in fact, likely welcomed with interest and a willingness to engage in dialogue about the philosophical dimensions of science.

However, there is something unique in Heisenberg's relationship with Cantore. In his exchanges with other Jesuits and philosophers, the focus remained on the intellectual exchange of ideas. In his correspondence with Cantore, Heisenberg encountered something entirely new, which often seems to catch him off guard: Cantore didn't want to just discuss the philosophy of quantum physics or merely address the gap between the two cultures. He wanted to found something concrete to solve the problem at its root. He saw it as part of his mission to establish an interdisciplinary institute at a university that would serve as a catalyst

---

34. This latter statement is included in Heisenberg, "Scientific Truth and Religious Truth." See also Heisenberg, *Ordnung der Wirklichkeit*; "Naturwissenschaftliche."

35. Weizsäcker, "Heisenbergs Entwicklung seit 1927," 40.

for dialogue between scientists and philosophers—at a time when the philosophy of science was just emerging, and with a strong "vocational" emphasis. This emphasis is evident in the scientific publications Cantore shared with Heisenberg, along with the drafts of memorandums in which the Jesuit laid out how the idea of scientific humanism could become a reality. To achieve this ideal, Cantore sought Heisenberg's help. Surprisingly, Heisenberg believed the idea had merit and should be pursued. He was willing to assist however he could, even crossing the ocean to visit Fordham University to contribute to the founding of the center Cantore envisioned, or lending his name to support the project. In one of his letters, Heisenberg wrote: "I had the impression that Dr. Cantore is one of the few people who can help to bridge the gap between modern science and the old humanistic philosophical tradition of Europe."[36]

Four Academic Visits and Meetings: Conversations on Scientific Humanism

We can only imagine Cantore's excitement upon receiving a letter from Heisenberg and reading the words addressed to him in German on the typewritten page: "Lieber Herr Cantore."

The continuation of nearly a decade of correspondence and at least four meetings in Munich, where the philosopher and the great physicist were able to converse (in the summers of 1967, 1968, and 1971 and October 1973), can only be explained by a shared goal—an ideal that both men held in common. Before delving into the details of this decade-long correspondence, it is worth reflecting on the four in-person meetings between Cantore and Heisenberg and asking what these encounters reveal about the relationship between the two intellectuals.

These meetings took place both in the professional setting of the Max Planck Institute, where Heisenberg, particularly during the last three visits, allowed Cantore to interact with several internationally renowned scientists attending the Institute's activities. The meetings also likely occurred in more personal and familial settings, as suggested by a letter after the 1971 visit, in which Cantore thanked Heisenberg's wife, Elizabeth, for her hospitality.[37]

---

36. Letter 28 from Werner Heisenberg to Detlev Wulf Bronk, Rockefeller University (April 1, 1968).

37. "The first motive for writing this letter is deeply felt gratitude. Our recent encounter has been for me a greatly enriching experience. The privilege of spending some

We know little about the content of their first conversation due to a gap in their correspondence between May and August 1967. However, from subsequent letters, it seems they discussed Heisenberg's philosophical contributions in *Physics and Philosophy* and a 1958 lecture on "Planck's Discovery and the Philosophical Problems of Modern Physics" from his text *On Modern Physics*, as well as Cantore's research, beginning with his doctoral studies on the philosophical questions of microphysics, and his future publication plans. In a letter where Cantore commented on Heisenberg's essay on Goethe and Newton's theories of light, it appears that Cantore also discussed his own essay on optics, published the previous year, for which he had promised Heisenberg an offprint.[38] At the end of his stay in the summer of 1968, following the bitter disappointment of being dismissed from Fordham, Cantore wrote gratefully to Heisenberg from Oxford: "My sojourn in Munich was a unique occasion for getting acquainted with science as a concretely lived activity. The result has been that my esteem for the scientists has increased immensely, and my determination to serve the cause of scientific humanism has been vastly strengthened" (Letter 41, November 7, 1968).

As for Heisenberg's demeanor during these meetings, Cantore's remarks in a circular letter to the members of the Institute for Scientific Humanism, written shortly after Heisenberg's death in 1976, offer insight: "The aspect of Heisenberg's humanistic message about science that struck me most in our private conversations has been responsiveness. He was in agreement that science is creative; however, in his view, the root of such creativity was nothing but a responsive attitude. Being a gifted amateur musician, he used to resort to a musical simile. The creative scientist—he contended—is nothing better than the ordinary person, except for one respect: he has a more sensitive ear than the average and responds more readily and perseveringly. This attitude of responsiveness may very well be taken as the key to understanding the humanistic and humanizing import of science" (Letter 106, June 15, 1976).

---

hours at your home with you and Mrs. Elisabeth has made me realize more concretely how science can acquire a thoroughly human significance, including family life. This experience was for me very heartening. I thank you and your wife for it." Letter 91 (September 21, 1971).

38. See Cantore, "Genetical Understanding of Science."

Ten Years of Letters: The Development of an Academic Friendship

When it comes to the tone of their correspondence, the matter is more complex and involves various factors, including how it evolved over the decade. Initially, the relationship was marked by a natural asymmetry, which later developed into a more equal partnership. Cantore was, at first, an early-career researcher addressing a luminary in physics, and Heisenberg maintained a cautious academic distance at the beginning, welcoming Cantore and responding generously. The early exchanges were formal and academic: appointments arranged through Heisenberg's secretary Giese, Cantore sending some essays and his doctoral dissertation with requests for feedback, and unsolicited comments on Heisenberg's writings. Over time, however, their relationship transformed significantly, developing into what could be called an academic friendship, marked by mutual respect, although Heisenberg always maintained the role of senior mentor. With their personal meetings, the sharing of common philosophical and humanistic interests, and Cantore's increasingly emotional and sometimes explicit requests for help, Heisenberg grew convinced—something he repeatedly expressed in letters—of his desire to support and encourage Cantore in the latter's mission to establish scientific humanism as both an academic and ideal pursuit, and also to translate it into a concrete initiative.

Taken as a whole, the letters reveal a narrative arc defined by moments of intense emotion, during which Cantore exposed his vulnerability and sought Heisenberg's help and support. The peak of this arc occurred in 1968, coinciding with the attempt to establish a specialized institute at Fordham University and Cantore's search for a stable academic position in the United States. This was a complex period in which Cantore juggled efforts to integrate into Fordham's highly structured academic environment while conserving his energy to launch the new institute. He submitted memoranda to university authorities, requested adequate space and financial support, and sought endorsements from internationally renowned scientists. The idealism and enthusiasm that led Cantore to request the conferment of an honorary doctorate on Heisenberg—which the physicist later declined—and to try to bring him to the Fordham campus collided with the realities of academic life, into which Cantore had perhaps entered with too much zeal and urgency. Research conducted in the Fordham archives reveals two perspectives on the matter: the academic institution, which recognized the merit of Cantore's initiative but saw his approach as premature and some of his requests as inappropriate (such as asking his superiors in Rome to assign him permanently to Fordham); and Cantore's perception, which blamed the situation on the institution's failure to understand the ideal of scientific humanism due to the cultural divide between the sciences and the humanities.[39] Those familiar with academic environments would likely agree that both sides had valid reasons. Fordham was going through a particularly difficult period, as evidenced by leadership changes shortly after Cantore's dismissal, backed by the Jesuit General Curia in Rome. However, Cantore was probably correct in perceiving a certain skepticism among the scientific community toward an innovative initiative that was marginal to the traditional academic discourse, as noted by both Heelan (who observed that Cantore's work could be considered as para-academic) and Heisenberg (who, after the 1970s, advised Cantore to propose his idea of scientific humanism in broader public intellectual contexts).[40]

---

39. See the appendix at the end of this book.

40. "Now, public recognition can only be acquired over a long period of time and over the years; even a single good book is usually not enough. Perhaps you should try even harder than you have been able to do so far to participate in public institutions of intellectual life, e.g., as a university lecturer, participant in conferences and discussions." Letter 92 (October 8, 1971)

Enrico Cantore

A moment of great emotional intensity came when Cantore, facing discouragement after his dismissal from Fordham and struggling to secure another academic position in New York, found himself in financial uncertainty. Heisenberg, with great discretion and diplomacy, sought to intervene and was ready to fly to New York to prevent the dismissal and unlock the establishment of the institute. When it became clear that the situation was irreversible, Heisenberg invited Cantore to Munich, but not before recommending him for a position at Rockefeller University (April 1968). This attempt failed (May 1969), followed by the rejection of Cantore as the English translator of *Der Teil und das Ganze* (June 1969).

Cantore's disappointment seemed less about the events themselves and more about the fear of being unable to fulfill his mission. Heisenberg wisely encouraged him not to abandon his dream of scientific humanism. Cantore responded in February 1969: "I feel confident and desirous to develop my contribution to scientific humanism. . . . I have no words to thank you for the immense encouragement that you have given to me. You know how discouraged I was when I left Fordham! I thank you also because I know that I can count on your support for the future." And a few months later: "Frankly, I do not know whether I would ever have found confidence to go on in the present work toward a scientific-humanistic synthesis had it not been for your continual understanding and open-hearted support. Of course, I am not able to thank you in any adequate way. As an expression of my esteem for you, I shall engage myself thoroughly in this work" (Letter 67, June 27, 1969).

Heisenberg's compassionate and wise encouragement, speaking from the experience of his stage in life and career, continued to sustain Cantore through the difficult process of publishing *Atomic Order* and later *Scientific Man*. Cantore was initially disappointed by the poor reception of *Atomic Order*, despite the cooperation of MIT Press in distributing a significant number of copies to university professors—a distribution that resulted in only a few perfunctory responses and no concrete opportunities (Letter 79, December 17, 1970). Heisenberg responded a few days later, encouraging Cantore not to lose heart and sharing his own experience, treating him almost as an equal. The candor of Heisenberg's comment is almost endearing: "Even if a book is ultimately very successful in this regard, its influence only becomes apparent very slowly over the years. It also seems that in America, it is not customary to express gratitude for books received. For my book on the unified field theory, which was also sent to many American colleagues, I received only a very few responses" (Letter 80, December 21, 1970).

By the end of 1970, Cantore was teaching part-time at Fordham's Lincoln Center (Letter 79, December 17, 1970), where Jesuit General Pedro Arrupe had granted him permission to return following a change in administration. The funding requests involving Heisenberg were met with lower-than-expected support, insufficient to publish *Scientific Man* in the short term, which was subsequently rejected by MIT Press. Additionally, due to budget constraints, Fordham had decided not to offer Cantore any teaching opportunities. Faced with financial uncertainty, despite a modest income from his pastoral work in Harlem, Cantore again confided in Heisenberg: "The double blow mentioned made me doubt for a moment whether it would be reasonable at all for me to continue in this almost impossible undertaking. But your repeated assurances in the past gave me new courage. Hence I decided not to abandon the course begun" (Letter 91, September 21, 1971). Heisenberg replied promptly: "Should you continue on your current path despite all resistance, or should you scale back or change your goals? ... I genuinely care about your ideas reaching broader audiences and because I believe you have something valuable to contribute to the intellectual development of our time" (Letter 92, October 8, 1971).

The situation seemed to take a positive turn starting on August 30, 1973, when Cantore excitedly and emotionally announced to Heisenberg the founding of the Institute for Scientific Humanism at Fordham, his appointment as director of the Institute, and the forthcoming publication of *Scientific Man*. These encouraging developments led to a brief but warm letter of congratulations from Heisenberg a few months later, on February 20, 1974 (Letter 103). Heisenberg's declining health likely delayed his response, and indeed, there would be no further letters, as his condition deteriorated until his death two years later, on February 1, 1976. The correspondence dwindled, and Cantore's final letters, in which he shared updates on the Institute and the publication of *Scientific Man*, went unanswered. The last letter, dated June 29, 1976, was one of condolences to Heisenberg's faithful secretary, Annemarie Giese (Letter 107).

Despite the challenges, we can consider the decade from 1967 to 1977 as one of the most productive periods for Cantore. The intellectual stimulation and experiences he encountered, combined with Heisenberg's unwavering support, led him to reach a peak in his academic output on scientific humanism during this time: two books and ten peer-reviewed articles in international journals, not to mention a considerable body of unpublished texts and memoirs.[41]

---

41. See *Published Writings* on p. 45.

Enrico Cantore

Although the fate of the Institute for Scientific Humanism at Fordham after 1985 remains unknown,[42] Cantore's mission took an unexpected turn in Rome two decades later. As Tanzella-Nitti's essay concludes, the idea of scientific humanism took concrete shape following their contact in the early 2000s with the founding of the International School for Interdisciplinary Research in Rome. This institute was inspired by the ideal foundations laid by Cantore's dialogue with the promoting group, led by Giuseppe Tanzella-Nitti, an astrophysicist and professor of fundamental theology at the Pontifical University of the Holy Cross.[43] The project to which Cantore had dedicated his life was not only realized with Heisenberg's help but continued in ways neither Cantore nor Heisenberg could have imagined, extending even beyond their lifetimes. After all, don't great ideas outlive us? As Angelo Secchi wisely wrote to McLaughlin: "At any rate, God has his own ways, and leads men willingly unwillingly along them" (April 13, 1968).

---

42. See *Epilogue* on p. 267.
43. Cantore, "Un Rapporto sul mio Apostolato."

2

Enrico Cantore: An Intellectual Profile

The Hidden Work of a Scientific Humanist

Giuseppe Tanzella-Nitti

Due to the unique circumstances surrounding his intellectual journey, the work of Enrico Cantore (1926–2014) remains relatively unknown today. Perhaps because it was both innovative and, in some ways, unconventional, his work did not consistently find a place within academic circles. Instead, it evolved largely through his personal commitment, which was both scientific and ecclesial, manifested in numerous essays and lectures, as well as through friendships and scientific collaborations. Cantore's thought is preserved in two significant volumes, *Atomic Order: An Introduction to the Philosophy of Microphysics* (1969) and *Scientific Man: The Humanistic Significance of Science* (1977), along with several dozen articles and a substantial number of unpublished works, including at least two additional volumes of several hundred pages each.

His teaching experiences at the Gregorian University in Rome and Fordham University in New York during the 1960s were too brief to establish a lasting academic school of thought. Consequently, Cantore worked independently and personally, yet he was far from an isolated researcher. Throughout his life, he had the opportunity to meet and collaborate with renowned scientists, such as Werner Heisenberg, who invited him to spend a period of study with him in Berlin. Cantore also maintained correspondence with scholars like Jean Piaget. For example, after publishing an essay in 1971 in the *Philosophy of Science* journal titled "Humanistic Significance of Science: Some Methodological Considerations," which explored the religious and existential dimensions of scientific inquiry, Cantore received letters from 150 scientists worldwide expressing interest in his work. The experience of founding the Institute for Scientific Humanism in New York in 1974, though limited in scope, further demonstrates the network of human and scientific relationships in which Cantore was engaged and upon which he had influence.

By contemporary standards, his work might be hastily classified under the philosophy of science. However, such a classification would be both narrow and incomplete. As a Catholic priest and member of the Society of Jesus, Enrico Cantore was a profound biblical scholar and a theologian deeply passionate about the mystery of Christ. His interdisciplinary approach extends beyond philosophy, starting from the sciences and reaching into theology. Cantore never limited his reflections to the epistemological plane but naturally transitioned to the anthropological level, examining the personal and social resonances of his ideas. His thinking about science was primarily a reflection on scientific activity—that is, on the work of the scientist, which he understood with rare depth, capturing its humanistic, existential, and religious dimensions, as well as its ethical and social implications.

To find a comparable vision of the scientific enterprise, one might look back to Maurice Blondel's *Action*, particularly where the French philosopher examines the motivations behind scientific research, to some of Teilhard de Chardin's essays, especially those discussing the engagement of scientists in the world and the church, or to the writings of Pavel Florenskij, though his hermeneutics are far from simple. Michael Polanyi's work, rich in insights into the personal and contextual nature of scientific knowledge, had redirected many authors' attention to the importance of the subject. Thomas F. Torrance developed a biblically founded interdisciplinarity that bridged the sciences and theology, while Karl Popper explored the relationship between science and society, including the social responsibility of scientists. Romano Guardini made significant contributions to understanding the relationship between humanity and technology, reflecting on the Christian sense of progress. Gilbert Simondon developed a philosophy of technology that sought to integrate it into human culture, avoiding a reductive view of technology as merely a source of alienation. It may be surprising, but it is true that Enrico Cantore's works contain all these components, presenting them in an organized and convincing manner. He orchestrates a dialogue between scientific thought, philosophy, religious sense, and theology, following an engaging phenomenological approach that avoids easy concordism or forced apologetics.

Enrico Cantore speaking at a conference
with Evandro Agazzi (second from the left), end of the 1960s.

Rather than being merely a philosophy of science, Cantore's work is a *philosophy of scientific and technical activity*, but also a *theology of nature* and a *theology of science*. The latter is understood as a reflection on the role that scientific knowledge and technical-scientific progress play in God's plans for creation and, consequently, in the history of salvation. In the 1960s, he developed the idea and content of a "scientific humanism," a concept recently revived by Bruno Latour, though Latour employs it with goals quite distant from the original biblical perspective of the Jesuit thinker, who clarified that his intention was to present a "sapiential scientific humanism." Cantore's work is strongly Christocentric, boldly associating the mystery of Christ with the most intimate aspects of nature and the highest aspirations of the scientist. The scientific humanism he proposes opens up a vision of the unity of knowledge that aims to courageously reevaluate the nature and mission of the university.[1]

An interesting reference to such an understanding of "scientific humanism" can be found in an address by Pope John Paul II to the Pontifical Academy of Sciences in 2000:

> [We refer to] *humanism in science* or *scientific humanism* in order to emphasize the importance of an integrated and complete culture capable of overcoming the separation of the humanistic disciplines and the experimental-scientific disciplines. If this separation is certainly advantageous at the analytical and methodological stage of any given research, it is rather less justified and not without dangers at the stage of synthesis, when the subject asks himself about the deepest motivations of his "doing research" and about the "human" consequences of the newly acquired knowledge, both at a personal level and at a collective and social level.[2]

Several sections of this speech address the humanistic dimensions of scientific research.

These are just a few of the reasons why Enrico Cantore's work deserves to be more widely known and continued, especially considering the significant questions raised by the scientific and technological era of the twenty-first century, many of which are centered on how to understand the relationship between scientific thought and humanism.

From 1999 onward, my personal memories of Father Cantore come into play. After reading his book *Scientific Man*, I felt compelled to correspond with him, inviting him to contribute an entry on "Scientific Humanism" for the *Dizionario Interdisciplinare di Scienza e Fede* (English version online: *Interdisciplinary Encyclopedia of Religion and Science*), which I was editing with Alberto Strumia. This work was later published in Rome in 2002. Cantore was a thinker of great depth but also complexity, and I recall having to ask him several times to clarify the meaning of some of his neologisms, as their translation into Italian was not always straightforward. In 2004, I had the opportunity to spend a period of study and engage in fruitful conversations with him in Oradell, United States. This marked the beginning of a more intense collaboration focused on two main objectives: first, to develop reflections that would help scientists appreciate the humanistic and philosophical-wisdom dimensions

---

1. See the foreword to Francis, "*Veritatis Gaudium.*"
2. John Paul II, "Address to the Pontifical Academy of Sciences."

of their work, and second, to foster greater interest within the Catholic Church toward the scientific community. As priests, we both recognized the need for a specific pastoral ministry aimed at scientists, and as theologians, we shared the idea of promoting a careful use of scientific findings in theological development.

Enrico Cantore with Giuseppe Tanzella-Nitti,
Berger School in New Jersey, July 2004

In the summer of 2006, Father Cantore returned to Italy, settling at the Jesuit Residence in Rome, on Via degli Astalli, where, despite his advanced age, he continued to study and write on the topics that had occupied him throughout his life. During his years in Rome, he followed and inspired an interdisciplinary research group that Alberto Strumia and I had initiated in 2005, aimed primarily at young graduates in scientific disciplines. This group later became part of the SISRI (an Italian acronym standing for "Scuola Internazionale Superiore per la Ricerca Interdisciplinare," in English "International Advanced School for Interdisciplinary Research"), established in 2013. During this period of his life, Cantore focused on drafting memoirs and notes with ideas and suggestions for the church's mission in the scientific community, following with interest the activities of SISRI, in which he saw the realization of the scientific and ecclesiastical ideals that had inspired the Institute for Scientific Humanism, though the latter was directed at senior scientists rather than young scholars in training.

In December 2012, due to deteriorating health and a diagnosis of kidney cancer, Cantore was transferred to the infirmary of the Society of Jesus in Gallarate. He passed away just over a year later, on March 27, 2014, the feast day of Blessed Francesco Faà di Bruno, a fellow Piedmontese priest and scientist. Always in good spirits, with a youthful demeanor, a deep knowledge of Scripture, and a modest and sober character, Enrico Cantore accepted with optimism and goodwill the fact that he had to work for most of his years almost in solitude—praying, writing, and studying, but also inspiring and encouraging those who came into contact with him.

Contribution of Cantore's Thought to the Current Debate on Scientific Humanism

Among Enrico Cantore's most notable contributions is his effort to bridge the "split between the two cultures," namely the scientific and the humanistic—a problem that Charles P. Snow highlighted long ago but whose roots can be traced back to Wilhelm Dilthey's distinction between the human and natural sciences.[3] While Snow, Guardini in his reflections on technology,[4] and John Henry Newman in *The Idea of a University* (1852) all proposed solutions that emphasized a holistic education to enable dialogue between the humanities and the sciences, Cantore's approach takes these ideas further.

Newman's ideal of the gentleman—an educated person equally versed in liberal arts, whether a scholar or a scientist—Snow's vision of a literate scientist familiar with the history of science, and Guardini's hope for a humanistic sensibility guiding the powerful engine of scientific and technological progress all resonate in Cantore's thought. Guardini's metaphor in *Letters from Lake Como* (1925) of a powerful motorboat being navigated with the sensitivity of sailing a sailboat captures the careful balance required in steering modern progress. Cantore, however, not only incorporates these perspectives but also transcends them. He argues

---

3. See Snow, *Two Cultures*.
4. See Guardini, *Letters from Lake Como*.

that scientific knowledge is itself a source of philosophical thought and a means of humanization. The humanistic perspective is not something external to be added to scientific endeavors; rather, it is inherent within the scientific activity itself.

Science places humans before themselves, prompting the same questions that inspired the birth of philosophical thought. Scientific knowledge enhances human dignity by revealing our place in the cosmos, showing what we are capable of, and calling us to exercise solidarity. These humanistic dimensions are not pursued in spite of science or the scientific mindset but emerge precisely from within science. Science, therefore, does not replace classical humanism with a pseudo-humanism fit for our times, but instead reveals the contribution of science to the one true humanism—a humanism expressed in the various manifestations of human culture and guiding the great philosophical questions about existence.

Cantore's perspective offers a more convincing approach to the relationship between science and humanism, especially in the context of contemporary debates, particularly after Hiroshima. Today, a neutral or functional view of science is prevalent. Science, in this view, is not seen as a source of humanistic thought but as a mere tool—neutral by nature—that humanity can use for good or ill. This perspective allows many to sidestep the ethical issues raised by science by separating the scientist from the applications (which they themselves often design). In this view, the direction and goals of scientific and technological progress should be decided in ethical-political spheres, for example, through a non-expert democratic consensus, essentially distinct from the activities of science and the scientist. This pragmatic, ultimately heteronomous orientation positions the scientist as a technical operator: knowing more, they are called to deliver more pragmatically.

In recent decades, a vision of science as a "third culture" has gained traction. This "new" humanism is constituted by science, which is called upon to solve problems and make decisions through the efficiency of the scientific method, effectively replacing philosophical reflection, which is deemed unreliable due to its lack of experimental basis. As a result, society would be led not by humanists or traditionally understood scientists, but by men of science promoted to the role of moral guides, now able to propose themselves as politicians and governors. The scientist, knowing more, could judge more and about more.

Cantore's *sapiential scientific humanism* offers a different perspective on the relationship between science and humanism. He asserts that science possesses an intrinsic humanistic dimension, serving as a source of moral dignity, educational resources, and freedom. Scientific knowledge is an experience of service and freedom: through it, we can, in principle, overcome or at least reduce the limitations of poverty, disease, and underdevelopment. It also allows us to free ourselves from the constraints of ignorance and superstition. From this standpoint, it is clear that the researcher, as a scientist, has a specific responsibility for social and cultural advancement: knowing more, they can and must serve more.

In today's socio-cultural context, putting Cantore's vision into practice is neither easy nor immediate, given the complex interplay of science, technology, and economics that we are bound by, as well as the fact that Cantore primarily focuses on basic science and only secondarily on technology. Nevertheless, it is evident—and this must be acknowledged—that only by transcending the neutral and pragmatic views of the scientific enterprise can we find the resources necessary to overcome the uncertainties of contemporary times.

Another outcome of Cantore's approach is that the dialogue between scientific and humanistic disciplines—or even the dialogue between science, philosophy, and theology, when the latter is called upon—can more easily shift from the epistemological level to the anthropological one. The epistemological level remains essential, as common themes cannot be discussed without clarifying the respective domains of knowledge, methods, and formal frameworks involved. However, this level is insufficient. It is only by reaching the anthropological level that we can address the most existentially significant questions because science, like any true knowledge, raises them as it confronts the world and ourselves. An authentic unity of knowledge can only be built around the intellectual unity of the knowing and reflecting subject, not by attempting to unify the object or the method.

It is evident that the debate between science, philosophy, and theology—consider the great questions about the origin of the cosmos and life—is currently centered mainly on the epistemological plane. This limitation is also present in theology, often leading to a flattening of the discourse on the biblical-hermeneutical level, with little ability to engage with questions of meaning, which are considered foreign to science, while epistemological and methodological questions seem to more easily bridge gaps. Indeed, questions of meaning are foreign to the scientific method when understood impersonally and objectively, but they are not foreign to the scientific activity as Cantore defines it. The researcher does not abandon the scientific method when moving to the anthropological plane; instead, they simply add something more, recognizing as significant questions that arise from scientific knowledge but that cannot be fully answered by the scientific-experimental method alone.

Once the scientific activity is acknowledged as having access to a philosophical-sapiential dimension intrinsic to the work of the scientist—not as a heteronomous reflection imposed by external subjects—it becomes easier to incorporate the religious dimension and, in a reflective and rigorously rational manner, the specific contributions of theology. For Cantore, the religious dimension is inherent in scientific work; indeed, the scientific work itself is accompanied by feelings that can be aptly described as characteristic of a religious sense. Cantore grounds this thesis in a historical-phenomenological—or as he would say, genetic—manner, allowing the protagonists to speak for themselves. It is through the quotations of numerous scientists with whom Cantore ideally dialogues that the reader is brought to recognize feelings of wonder, reverence, and veneration, paralleling religious experience. Especially in his dialogue with Planck, Heisenberg, and Einstein (the three scientists most frequently cited by the Jesuit philosopher), Cantore speaks of the "shock of the unexpected" and the "experience of the foundations" as the scientific experiences that most profoundly express the encounter with an Otherness, the listening to the Absolute, and the contact with the foundation of intelligibility and order.[5]

In particular, Cantore believes that the connection between the religious sense and scientific activity is founded on two key elements: those engaged in scientific research a) experience the "foundations" (*experience of ultimates*), and b) exhibit reverence for the transcendent mystery present in nature, with an intensity and depth inaccessible to others. This

---

5. "We can understand the justification of the view that intimately connects science with religion. The reasons that support it are essentially two. In the first place, consistently interiorized science amounts to an original experience of ultimates. Science makes man perceive the wonder and awe of nature in a way that is inaccessible to the nonscientific person. In the second place, the interiorization of science is capable of making the reflective person perceive—at least vaguely—the very source of existence, which accounts ultimately for the intelligibility and wonderfulness that science continually detects in nature. Science gives to attentive and responsive man a new sensibility for the transcending mysteriousness manifested by the overwhelming richness and beauty and power of nature, which science alone is able to reveal to man." Cantore, *Scientific Man*, 131.

religious-existential phenomenology can easily interact with theological reflection, which can now reveal the personal nature of the Foundation, the Absolute, and the filial nature of the Logos-Word that presides over the mystery and meaning of the cosmos, showing that what is "given" in science can be recognized by the scientist as "gifted." In this way, the answers provided by Judeo-Christian Revelation can enter into a close and spontaneous connection with the questions raised by scientific activity, without forcing or concordism, starting from what scientific research, as understood by Cantore in a personalistic and existentially engaging way, is capable of indicating through its protagonists.

Ultimately, it is the origin from the Creator and the intimate and transcendent meaning that nature holds that explain why scientific research is so compelling, why it can become a life's passion, and why doing science is "a labor of love." As Cantore clearly states, "Science is love because it enables man to establish a relationship of personal communion with nature and with nature's own ultimate source of meaning."[6]

The Legacy of Enrico Cantore

Cantore's ideas extend beyond theoretical considerations to encompass practical applications as well. His assertion that science is a source of humanization—for both the scientist engaged in research and society as a whole—implies a dual transformative effect. On one hand, it enhances and reveals the virtues of the scientist, ultimately contributing to their personal dignity. On the other hand, it influences society at large, suggesting that only through a science conducted with a humanistic perspective can scientific and technological progress truly translate into human advancement.

These are undoubtedly demanding ideals, which must contend with the complex interplay between science, economics, and politics that exists today. They are challenging, if not impossible, to realize if left to the goodwill of individual researchers alone. However, these ideals are crucial in pointing the way forward. At the end of many of his public lectures, Enrico Cantore was often told that he held too "high" an ideal of science. This observation, while not unfounded and seemingly congruent with reality, is not a critique of his thought, nor does it judge it as weak or baseless. It is similar to the comments one might make about Thomas

---

6. Cantore, *Scientific Man*, 143.

Aquinas regarding his lofty conception of reason or truth. Aquinas held a high view of human reason, likely not shared by many of his contemporaries, yet that did not detract from it being the true idea of human reason—capable of unifying humanity, exposing error and contradiction, and ascending from creatures to the Creator.

Similarly, the idea that Enrico Cantore had of scientific activity might seem idyllic or reserved for a select few. Yet, it is precisely this vision of science that justifies why a young person feels drawn to scientific research, why a seasoned scientist is capable of persevering with sacrifice in their studies and inquiries, or why society looks to scientists for certain solutions and behaviors.

This leads to Cantore's reflections on the social responsibility of scientists. Cantore envisioned scientists as cultural leaders, capable of laying the groundwork for genuine social advancement. Unlike the concept of the "third culture," the scientist Cantore envisages is not a politician governing society through the scientific method or by relying on sophisticated artificial intelligence algorithms. Instead, Cantore highlights the scientist's ability to deeply understand reality, to be more awed by it, to observe nature with greater respect as something entrusted to us, and to grasp more clearly the uniqueness of humanity in the cosmos and its specific dignity. The scientist helps us understand what truly makes us human and what corrupts us. For Cantore, the leadership of men and women of science is a cultural leadership, not a political or ideological one, though it undoubtedly serves as a source of political and social thought.

In his 1979 lecture "Leadership for Human Dignity: The Developmental Challenge to Scientific Professionals," Cantore asserts that scientists are the natural leaders of development due to their expertise and social influence. To fulfill this leadership role, scientists must not evaluate everything through a scientistic reductionism but should instead convey the idea of scientific knowledge as a factor in human dignity and the foundation for all progress, including human progress. This does not mean abandoning their method, but rather fully utilizing it, following the philosophical insights it suggests: "Scientific professionals," Cantore clarifies, "to be developmental leaders, far from weakening their fidelity to their profession, should increase it."[7]

---

7. Cantore, "Leadership for Human Dignity," 252.

For Cantore, a Jesuit priest and believer, speaking of humanism means speaking of Jesus Christ. His Christological and Christocentric vision seems to serve three main purposes. First, the Christological reference allows him to frame the reflection on intelligibility and order in nature within a reflection on the Christian Logos, a personal logos, thus indicating the appropriate path to overcome pantheism or deism—what we might today call naturalism—that is closed to transcendence. These views are not uncommon when scientists reason about the laws of nature and the foundation of the world.

Second, scientific and technological progress is not an immanent practice aimed solely at improving human living conditions, but rather a participation in the ongoing construction of creation, a construction possible only in and through Christ. United with Christ, humans can, as Cantore would say, engage in "quasi-creativity."

Finally, Jesus Christ, true God and true man, is the model of true humanism, which scientific humanism must also express. The Paschal mystery of Christ, with its law of charity and service, is the only source of energy and grace capable of sustaining the sacrifice that the pursuit of truth entails, even in scientific work, and orienting it toward altruism and the common good.

The biblical foundation of Enrico Cantore's thought is primarily drawn from the wisdom books. He intelligently parallels the figure of the sage with that of the scientific researcher, observing that the latter, modeled after the former, must become capable of listening to God through the study of nature and is called to cultivate the same virtues: diligence, care, humility, attention, and reflection on the philosophical questions that arise from observing reality. We are not far from the truth—and the chronological sequence of the articles presented in this volume demonstrates this—when we say that the fundamental intuition of a *sapiential scientific humanism*, with all its previously discussed implications, was born precisely from careful meditation on these scriptural texts. Here also lies the foundation for the association of Christ with Wisdom, a connection Cantore frequently employs, offering a complementary vision to what Eastern theology would do, which tends to emphasize the trinitarian dimension of Sophia without contradicting it.

It might be worthwhile to explore the points of agreement and differences between Cantore's Christocentrism and that of other authors who were attentive to scientific discourse, such as Pierre Teilhard de Chardin—a thinker with whom Cantore's writings do not directly engage—as well as Thomas F. Torrance and Denis Edwards, who are more contemporary.

What could Cantore's thought contribute to today's cultural climate, where scientific culture and technological applications are exerting increasing influence in twenty-first-century society? And what might he have to say, more generally, to the contemporary Science and Theology movement, which aims to intensify the dialogue between theologians and scientists? Anyone who approaches the writings of this Jesuit from Piedmont will easily see that his contribution would undoubtedly be significant, but it requires proper exposure and systematic illustration. Consider, for instance, the uncertainties surrounding the relationship between humanity and technology, between seeking spaces for the human and technological society—a relationship that is often viewed as conflictual, sometimes even within ecclesial circles. When discussing scientific and technological progress, the call for caution is usually the first to be voiced, rather than a discourse on the humanizing potential of scientific activity.

Moreover, imagine the shift that philosophy and science could undergo if, inspired by Cantore's vision, they were encouraged to move from the "critique of knowledge," where they have been largely confined throughout much of the last century (and into the present), to a genuine "philosophy of scientific and technical action," which could finally highlight the role of the subject and the personalist dimensions of knowledge. The ethical gains would be considerable, as the ethical-philosophical dimension of research, in Cantore's view, would be connected to the scientist's reflection on the object and goals of their activity, rather than being externally imposed by other subjects.

This approach would also protect the freedom of research itself, distancing it from the functionalist and instrumental model of science that still predominates today: science is free because its activity is intrinsically oriented toward truth and the good, conducted by personal subjects and therefore associated with corresponding responsibility. When science understands itself as a neutral instrument, it easily falls prey to the seductions of economic, political, and ideological power.

Reflecting on the philosophical dimensions of scientific knowledge, as Cantore does, including the religious experience that such knowledge can evoke, ultimately helps restore unity to reality. It acknowledges hierarchically ordered levels that allow the subject to access different levels of inquiry and meaning. It also strengthens the unity of the intellectual experience of the knowing subject, who rediscovers the possibility of being a scientist, philosopher, and existentially engaged person (on a religious level), without causing methodological rigor to falter or creating uncomfortable intellectual divides.

I hope that a more thorough understanding of this author will encourage renewed reflection on the problems he addresses, an examination of the solutions he proposes, and a dialogue with other thinkers who have advanced similar perspectives and demonstrated shared sensibilities. This will sometimes require going against the grain, exposing oneself to the critique of holding too high or idealistic a view of science. However, it is not by downplaying this vision that the challenges of the scientific and technological society of the present century can be addressed, but by clearly indicating the direction in which we must move, even if it is demanding.

3

Enrico Cantore: Biography, Chronology, and Publications

A Biographical Sketch

ENRICO CANTORE, THE ELDEST of six children, was born in San Mauro Torinese on July 19, 1926. He entered the Society of Jesus in Cuneo on August 31, 1943, where he remained until 1948. He then pursued his philosophical studies at the Aloisianum, the Jesuit center in Gallarate, completing a three-year program. In 1951, Cantore continued his theological education in Turin, while simultaneously enrolling in the Faculty of Mathematics and Physics, from which he graduated four years later.[1]

In 1955, Cantore moved to Chieri to obtain a licentiate in theology at the Jesuit theological institute "Casa Sant'Antonio." During this time, he developed a deep interest in Sacred Scripture, particularly in the figure of the sage as depicted in Israel's wisdom literature. Cantore was ordained a priest in Chieri on July 13, 1958. In 1960, he completed the Society of Jesus' special year of probation in Austria, under the guidance of Father Peter Heimejer.

In 1961, Enrico Cantore began teaching at the Pontifical Gregorian University in Rome while also dedicating himself to the study of the

---

1. Father Enrico Cantore was enrolled at the University of Turin in the academic year 1951–1952 and graduated on July 9, 1955, with a thesis titled "The Philosophical Problems Arising from Quantum Mechanics." The graduation report mentions two sub-theses: "Unstable Particles in Cosmic Radiation" and "The Betatron."

philosophy of science.² His journey then took him to the United States, where he continued his work in philosophy of science at Santa Clara University in California (1962–1963) and later in Chicago (1963–1964). In 1964, the Society of Jesus assigned him back to Rome to teach at the Gregorian University, where he completed his advanced degree in philosophy. Shortly afterward, he decided to return to the US, where he established significant connections with researchers in the scientific community, gradually developing the idea of deepening and promoting a humanistic reflection on scientific activity.

In 1967, Cantore began teaching philosophy of science at Fordham University, focusing on the philosophy of quantum mechanics and preparing his book *Atomic Order: An Introduction to the Philosophy of Microphysics*, which was published by MIT Press in 1969. He spent two years in Europe (England, Germany, Italy), including a period of research alongside Werner Heisenberg, at the latter's invitation. Upon his return to the US in 1969, Cantore settled in New York, where he would remain until 1992. Initially, he resided in a Jesuit student house and later served as a chaplain at a high school run by the Christian Brothers in the Bronx.

In 1974, Cantore, along with other intellectuals, founded the Institute for Scientific Humanism in New York, and in 1977, he published *Scientific Man: The Humanistic Significance of Science*. Despite his significant efforts in establishing the Institute, the initiative was forced to discontinue a few years later due to a lack of resources and, perhaps, an insufficiently structured organization. In 1980, Cantore spent several months in India delivering lectures organized by the Jesuits. In 1992, he moved to Oradell, New Jersey, where he served as chaplain at Bergen Catholic School until the summer of 2006. During this time, Cantore maintained connections with scientists, especially in the United States, and continued to reflect on themes related to scientific humanism, particularly in historical and epistemological contexts.

---

2. He completed his doctorate in 1966 with a dissertation titled "Atomic Order: An Introduction to the Philosophy of Microphysics." An excerpt of his doctoral dissertation in philosophy is housed at the Gregorian University. The university's archives indicate that Cantore was enrolled in the doctoral program in 1961/1962 and defended his dissertation on June 18, 1966 (committee: Filippo Selvaggi, SJ, and Mario Viganò, SJ).

Enrico Cantore

These years marked a period of growing concern for Cantore regarding the church's mission in relation to the scientific community. He dedicated himself to preparing writings, memoirs, and lectures aimed at guiding Christian intellectual outreach and emphasizing the role of scientists, particularly their openness to transcendence, which he believed was inherent in genuinely and ideologically unbiased scientific activity. Years earlier, the Piedmontese Jesuit had involved two Carmelite monasteries in his mission to evangelize the scientific culture, entrusting his intellectual apostolate to their prayers. One monastery was in Valmadonna, near Alessandria, where his sister, Sister Maria Angelica (born Giuseppina), resided, and the other was in Quart, near Aosta—a new foundation blessed by Pope John Paul II in 1989, where some of the nuns from Valmadonna had relocated.

Father Enrico Cantore returned to Italy in the summer of 2006. In December 2012, facing declining health and a diagnosis of kidney cancer, he was moved to the infirmary of the Society of Jesus in Gallarate, where he passed away on March 27, 2014.

Chronology

1926 (July 19)	Enrico Cantore was born in San Mauro Torinese
1943	Enters the Society of Jesus in Cuneo
1945	Still in his novitiate, at nineteen, feels God has entrusted him with a mission for Scientific Humanism
1948	Moves to Gallarate and studies Philosophy (three years)
1951	Theological studies in Turin. Begins Mathematics and Physics (1951/1952) at the University of Turin.
1955 (July 9)	Degree in physics, University of Turin. Moves to Chieri, biblical studies, especially on wisdom literature and the figure of the sage in the Sacred Scriptures.
1958 (July 13)	Ordained a priest in Chieri. First Mass in the Carmel of Valmadonna
1960	Probation year in Austria with Father Peter Heimejer
1961	Begins teaching at the Pontifical Gregorian University in Rome. Enrolled in the doctoral program in Philosophy (AY 1961/1962).
1962–1963	University of California Santa Clara, course in philosophy of science
1963–1964	University of Chicago, course in philosophy of science
1964	Destined to teaching at the Pontifical Gregorian University in Rome

1966 (June 18)	Concludes the doctorate in philosophy at the Gregorian with a dissertation titled "Atomic Order: An Introduction to the Philosophy of Microphysics." Committee: Filippo Selvaggi, SJ, and Mario Viganò, SJ.
1967	Teaches philosophy of science at Fordham University in New York.
1967–1969	Research stay in Europe (UK, Germany, Italy) with Werner Heisenberg
1969	Back in New York. Publication of *Atomic Order* (MIT Press)
1974	Founds the Institute for Scientific Humanism (ISH) in New York
1977	Publication of *Scientific Man* (ISH)
1977	Concludes a manuscript, *Science and Dignity* (unpublished)
1979	Upon suggestion of his colleagues, adds the word "World" to the ISH, thus becoming WISH
1980–1981	Conference travels to India
1985	Begins working (almost full-time) on a new book, *Christ Wisdom and Science*
1992	Moves to Oradell (NJ) as chaplain of Berger Catholic School
1999	Begins collaboration with Giuseppe Tanzella-Nitti
2006 (July)	Finishes typescript of *Christ Wisdom and Science* (unpublished)
2006 (September)	Moves back to Rome, Residenza del Gesù
2012	Diagnosed with kidney cancer, moves to Gallarate
2014 (March 27)	Dies in Gallarate, at age eighty-eight.

Published Writings (in chronological order)

1. "Philosophy in Atomic Physics: Complementarity." *Modern Schoolman* 34 (1957) 79–104.
2. "La soluzione fisica dell'enigma quantistico." *Divus Thomas* 60 (1957) 150–59.
3. "La sapienza biblica, ideale religioso del credente. I. Analisi del concetto ebraico di sapienza." *Rivista Biblica* 8 (1960) 1–9.
4. "La sapienza biblica, ideale religioso del credente. II. Aspetti intellettuali della sapienza biblica e loro evoluzione." *Rivista Biblica* 8 (1960) 129–43.
5. "La sapienza biblica, ideale religioso del credente. III. La sapienza biblica come conoscenza e timore-amore di Dio." *Rivista Biblica* 8 (1960) 193–205.
6. "Genetical Understanding of Science: Some Considerations About Optics." *Archives Internationales d'Histoire des Sciences* 19 (1966) 333–63.
7. "Some Reflections on Man's Unending Quest for Understanding." *Dialectica* 22 (1968) 132–66.
8. "Scientific Humanism and the University." *Thought* 43 (1968) 409–28.
9. *Atomic Order: An Introduction to the Philosophy of Microphysics.* Cambridge, MA: MIT Press, 1969.
10. "Science and Humanism: The Sapiential Role of Philosophy." *Dialectica* 24 (1970) 215–41.
11. "The Italian Philosophical Encyclopedia." *Review of Metaphysics* 24 (1971) 510–32.
12. "Humanistic Significance of Science: Some Methodological Considerations." *Philosophy of Science* 38 (1971) 395–412.
13. "Science as Dialogical Humanizing Process: Highlights of a Vocation." *Dialectica* 25 (1971) 293–316.
14. "The Humanity of Science: A Philosophical Response to a Thought-Provoking Essay." *Atti della Fondazione Giorgio Ronchi* 28 (1973) 165–86.

15. "La scienza e l'uomo. Significato della crisi umanistica contemporanea." *Civiltà Cattolica* 125 (1974) 112–29.

16. "Per una integrazione umanizzante tra scienza e uomo." *Civiltà Cattolica* 125 (1974) 322–36.

17. *Scientific Man: The Humanistic Significance of Science.* New York: ISH, 1977.

18. "Leadership for Human Dignity: The Developmental Challenge to Scientific Professionals." In *Issues of Development: Towards a New Role for Science and Technology,* edited by Maurice Goldsmith and Alexander King, 247–57. Oxford: Pergamon, 1979.

19. "Human Dignity, Science and Growth: The Humanistic Nature of Development." *WISH Basic Papers* 1 (1981) 1–24.

20. "Science for Human Dignity: A Christian Leadership Task." *Homiletic and Pastoral Review* 82 (1982) 47–53.

21. "Science, Religion and Human Dignity: Notes for Contemporary Evangelization." *Euntes Docete* 35 (1982) 213–24.

22. "La scienza come fattore umanistico." Translated by G. Tassani and S. Marmi. *Il Regno-attualità* 10 (1982) 216–19.

23. "The Christic Origination of Science." *Journal of the American Scientific Association* 37 (1985) 211–22.

24. "Umanesimo Scientifico." In vol. 1 of *Dizionario Interdisciplinare di Scienza e Fede,* edited by Giuseppe Tanzella-Nitti and Alberto Strumia, 1399–409. Rome: Urbaniana University-Città Nuova, 2002.

PART II

Correspondence (1967–1976)

I.

"I have taken the courage to write to you."

First Encounters, May 1967–October 1967

THE FIRST LETTER THAT Enrico Cantore wrote to Werner Heisenberg was posted on May 16, 1967. It began with Cantore introducing himself to Heisenberg, with the recommendation of Piero Salviucci, a prominent member of the Pontifical Academy of Sciences.[1] The Jesuit philosopher was forty-one years old at the time. Just a few months previously, he had moved to the USA, where he had been appointed assistant professor of philosophy of science at Fordham University in New York City.

The first meeting of Heisenberg and Cantore occurred in Munich, from June 15 to June 23, 1967. While unreported, the subject of their discussion can be easily inferred from the correspondence that follows the meeting. Heisenberg strongly advocated Cantore's concern about the philosophical importance of scientific practice, as outlined in "Genetical Understanding of Science: Some Considerations About Optics." This article, published the previous year, discussed the development of optics from a philosophical point of view, a subject of great interest to Heisenberg and one to which he had devoted a paper, "The Teachings of Goethe and Newton on Colour in the Light of Modern Physics."

---

1. Piero Salviucci (1899–1982) was former chancellor of the Pontifical Academy of the Sciences (PAS); nominated honorary academician in 1982. Werner Heisenberg had been nominated as a member of the PAS on April 5, 1955 (twelve years before). Schrödinger was nominated as a member of the PAS in 1936.

In the following correspondence, Cantore sent Heisenberg an offprint of his paper to discuss the philosophical questions that had been occupying their minds during the meeting in Munich. The mood of Heisenberg's reply was one of genuine interest and respect towards the intellectual efforts of Cantore. Heisenberg's regard for Cantore's work is apparent from his willingness to engage with the manuscript Cantore was working on, a topic at the heart of his scientific and philosophical inquiries. Heisenberg's initial response to the Cantore manuscript was optimistic, saying that Cantore had "rightly identified the order of matter." This compliment on the part of a scientist of Heisenberg's eminence was, without doubt, a considerable spur to Cantore, who was, after all, at an early stage of his academic life. Heisenberg's willingness to be more in touch and his suggestion for Cantore to return to Munich to continue their conversation pointed out the respect and mutual intellectual interest each of them had developed for the other.

Cantore responded to this encouragement by Heisenberg with enthusiasm and deep gratitude. Indeed, recognizing the value of Heisenberg's mentorship, Cantore sought further advice, particularly in publishing. He asked Heisenberg to help him find his way through the publication process of his manuscript (*Atomic Order*) and even asked if Heisenberg would write a foreword for the book. Heisenberg declined, explaining that it was his general policy to refrain from writing forewords, even for colleagues. However, he did offer to help in another way: he would use his influence to write on Cantore's behalf to a publisher to expedite the publication process, a gesture of kindness that must have particularly gratified Cantore.

With impeccable insight, Heisenberg recommended that he immediately see Ruth Nanda Anshen, editor of the *World Perspectives* series at Harper & Row in New York.[2] Anshen was well-known for her work in bridging the gap between science and the humanities, and her interest in publishing works with this nexus made her an excellent liaison for

---

2. Ruth Nanda Anshen (1900–2003) was an influential American philosopher, writer, and editor. She was a student of Alfred North Whitehead at Boston University, where she cultivated a passion for bringing together scholars from diverse disciplines worldwide. She became a fellow of the Royal Society of Arts in London and was actively involved in organizations such as the American Philosophical Association, the History of Science Society, and the Metaphysical Society of America. Anshen is also recognized for her role as the editor of several book series, including the *World Perspectives* series, which was published by Harper & Row. The Ruth Nanda Anshen papers, 1938–1986, are preserved at Columbia University Library.

Cantore. The fact that Heisenberg pointed the manuscript in Anshen's direction showed he was informed about the publishing landscape.

This support became frequent in subsequent letters from Heisenberg to Cantore, who always expressed his reverential gratitude. He apprised Heisenberg about the status of his manuscript—regarding the views he had sought from Fordham University Press and the favorable response he had received from Ruth Nanda Anshen. The enthusiasm of Cantore was very much visible as he suggested his desire to keep their intellectual exchange alive. He suggested that Heisenberg visit Fordham University campus to see the University's efforts at dialogue between the scientific and humanistic cultures—a theme that ran directly through their correspondence.

Letter 1. Pietro Salviucci to Werner Heisenberg (May 4, 1967)

Pontifical Academy of the Sciences,
the Chancellor

Vatican City: May 4, 1967

Your Excellency,

I would like to introduce to you Father Enrico Cantore, whom I have known for several years. Father Cantore, a member of the Society of Jesus, is a professor of the philosophy of natural sciences at the Pontifical Gregorian University in Rome.

He has long been particularly interested in the philosophical questions surrounding atomic physics.

As is evident from Father Cantore's accompanying letter, I fully understand how much he would value the opportunity to have a conversation with you.[3]

Consequently, I would kindly ask you to let him know if and when he might visit you at your residence in Munich.

With my best regards and sincere thanks, I remain,
Yours respectfully,

(Prof. Dr. Pietro Salviucci)

His Excellency
Professor Dr. Werner Heisenberg
Member of the Pontifical Academy of Sciences
Rheilandstrasse 1, Tübingen.

---

3. It refers, most likely, to Letter 2 (May 2, 1967).

Letter 2. Enrico Cantore to Werner Heisenberg
(May 2, 1967)

Pontifical Gregorian University
Piazza della Pilotta 4, Roma

May 2, 1967

Dear Professor Heisenberg,

As Professor Salviucci mentioned in his accompanying letter, it would be a great honor and pleasure for me if it were possible for you to grant me a meeting.

    I am particularly interested in the philosophical questions that arise from microphysics. This area of study has occupied me for many years, and I have even attempted to develop a genetic-philosophical method to address such questions in the best possible way and, if feasible, to solve them philosophically. It is precisely regarding such a method that I would very much like to speak with you. For you embody a rare combination of expertise: you have not only been instrumental in shaping modern atomic physics but have also profoundly and repeatedly discussed its philosophical implications. Thus, it would be exceptionally promising for me to hear your perspective in person.

    During the upcoming month of June, I will very likely have the opportunity to travel to Germany and Munich. I would be most grateful if you could let me know whether you might be available during that time for a meeting. Unfortunately, I am currently unable to provide exact dates. However, if you could inform me of the times when you might be reachable in Munich, I will notify you as soon as possible of the precise date of my arrival.

Hoping to meet you in person soon, I thank you sincerely.
Yours respectfully,

(Enrico Cantore, SJ)

Letter 3. From Annemarie Giese to Enrico Cantore (May 8, 1967)

May 8, 1967

—Secretary—

To His Reverend Excellency
Father Enrico Cantore
Pontificia Università Gregoriana
Piazza della Pilotta, 4
Rome, Italy

Dear Reverend Father,

In the absence of Professor Heisenberg, who will not return to Munich from an extended trip abroad until after Pentecost, I take the liberty of acknowledging receipt of your letter dated May 2 with gratitude.

Allow me to add that, according to Professor Heisenberg's schedule, the period between June 15 and June 23 would be most suitable for the meeting you requested. Professor Heisenberg will be traveling for a week at the beginning of June and has other commitments outside Munich at the end of the month.

I hope that a date within the timeframe of June 15 to June 23 will also be convenient for you.

With best regards,
Yours sincerely,

A. Giese
Secretary

Letter 4. From Enrico Cantore to Annemarie Giese (May 16, 1967)

Pontifical Gregorian University
Piazza della Pilotta 4, Roma

Miss Annemarie Giese
Max Planck Institute for Physics
Secretary—Institut für Physik
8 Munich 23
Föhringer Ring 6

May 16, 1967

Dear Miss Annemarie,

Thank you very much for your kind letter of the 8th. I am grateful that I will have the opportunity to visit Professor Heisenberg in June. The suggested period, from June 15 to 25, is convenient.

Due to other commitments, I plan to arrive in Munich on June 14 and would like to meet with the Professor on the 15th or 16th. If possible, I would prefer to depart in the afternoon of the 16th or early on the 17th. I intend to stay with the Jesuits at Kaulbachstraße 31a.

I hope it will be possible to meet you in person and express my gratitude directly. Please let me know when and where it would be best to meet with the Professor. I will be traveling from my current address in two weeks.

With all best regards,
Yours

(Enrico Cantore, SJ)

Letter 5. From Enrico Cantore to Werner Heisenberg (August 4, 1967)

Fordham University, Department of Philosophy
Bronx, NY 10458

August 4, 1967

Dear Professor Heisenberg,

A few days ago, I was finally able to send you the promised offprint. It is merely an example focusing on optics.[4] However, I hope that this paper can illustrate, at least in part, the method that I consider most suitable for a philosophical understanding of the natural sciences. Indeed, it seems to me that only a systematic philosophical investigation—based on the history of ideas and the psychological genesis of key concepts—can bring to light the profound human significance of the natural sciences.

I greatly enjoyed reading your lecture on Goethe.[5] The great poet may have acted incorrectly in refusing to accept the results of Newtonian research. But, as you so aptly emphasized, he was right in being unable to separate truth from the concept of value. For him, Newtonian science—and, indeed, the scientific worldview as a whole—represented an understanding that left no room for the "unum, bonum, verum." Thus, Goethe dramatized in his own person the dilemma of the "two cultures." This, too, is the great philosophical problem we face today.

Although I have not yet published anything of significant value in this area, I dare to briefly express a few ideas that might contribute to solving this question. As I may have mentioned to you in person, I had to spend many years without publishing much, as I was seeking a promising path not shaped by the biases of particular philosophical schools. Now, as a result of my studies on atomic order, additional reflections, and some conversations with creative scientists, it seems to me that I have found a path that could lead to a synthesis.

The essential point, I believe, is to view the natural sciences as they truly are—namely, as a new, deeper way for humanity to perceive and respond to reality as a whole. Thus, it seems to me that the natural

---

4. The mentioned offprint is Cantore, "Genetical Understanding of Science."
5. Cantore is referring to Heisenberg, "Teachings of Goethe and Newton." Cantore was probably reading Heisenberg's philosophical writings. See *Contextualizing the Correspondence* on pp. 13, 19.

sciences should be recognized as a novel, comprehensive, and transformative human experience. Acknowledging this experience could allow modern humans to systematically assimilate it, leading to a deeper and better understanding of the world, humanity, and God. In other words, this path could pave the way for developing a form of humanism better suited to contemporary times.

The question remains: How can we study the natural sciences so that they emerge as a total human experience—both for individuals and members of society? I think this could be achieved by applying the genetic method. Traditional philosophical methods have proven inadequate because they have neglected a crucial factor: the constant dynamism of both human knowledge and the material world. Thus, by systematically investigating the origin and development of the natural sciences, it would likely be possible to arrive at a satisfying philosophical interpretation of them.

Undoubtedly, such a study would be challenging but worthwhile. For if the scientific mentality is not humanistically integrated, it can cause great harm; yet if properly understood, it can contribute greatly to human well-being. In particular, the scientific spirit can serve humanity's continuous progress. After all, is it not the natural sciences that have taught us how to achieve steady advancement? Progress means preserving what has been achieved while perfecting it through new discoveries and realizations. Therefore, it does not seem too bold to conclude that the desired integration could lead to a synthesis that embraces the values of the past while keeping the door open to continual improvement.

I am unsure whether I have expressed myself clearly. I am already deeply indebted to you for the meeting you granted me and for your promise to review my manuscript on atomic order. Now, I am also grateful for your patience in reading this letter. In any case, I simply wanted to convey my willingness to dedicate myself to solving the problems that are so dear to you. I hope to write a detailed plan for this philosophical research in the coming months. Of course, I would be delighted to send you a copy if I may.

Meanwhile, I thank you once again and remain, with all best regards,
Yours sincerely,

(Enrico Cantore)

Letter 6. From Enrico Cantore to Annemarie Giese
(August 4, 1967)

Fordham University, Department of Philosophy
Bronx, NY 10458

August 4, 1967

Dear Miss Annemarie,

As you may already know, I recently sent an offprint and a letter to Professor Heisenberg.[6] Now, I would like to address a few words directly to you.

I do not want to let this opportunity pass without expressing my gratitude to you in writing. You are one of those people who are too often overlooked because they conscientiously work in the background—and yet, you are indispensable for ensuring that everything runs smoothly. I want to let you know how much I owe you. My conversation with Professor Heisenberg was a great experience for me, and you prepared everything so well! May the Lord reward you richly for it!

Since I hold Professor Heisenberg's philosophical reflections in high esteem, I would like to ask if you could send me any recent offprints of his, should they be available. The most recent works I am familiar with are *Physics and Philosophy* and the lecture included in the book *On Modern Physics*.[7] I would also like to know how the latest edition of *Wandlungen* differs from earlier ones.[8]

Once again, many thanks for everything, and best regards. Please also extend my gratitude and greetings to your colleague.

Yours sincerely,
(Enrico Cantore)

Miss Annemarie Giese
Max Planck Institute for Physics
8 Munich 23
Föhringer Ring 6
Germany

---

6. See Letter 5 (August 4, 1967).
7. See Heisenberg, *Physics and Philosophy*; "Planck's Discovery."
8. See Heisenberg, *Philosophic Problems of Nuclear Science*.

Letter 7. From Werner Heisenberg to Enrico Cantore (August 16, 1967)

August 16, 1967

To His Reverend Excellency
Father Enrico Cantore
Fordham University
Fordham Road
Bronx, NY 10458 USA

Dear Mr. Cantore,

I apologize for not writing to you until now about your philosophical book, but I wanted to study your book thoroughly before answering you, and I have done so in the meantime. First, let me say in general that I found your book, especially the second part, truly excellent, and I wholeheartedly encourage you to publish it.[9]

You have rightly identified the order of matter—that is, the visible operation of organizing forces in material phenomena—as the central theme of modern natural science. Moreover, you have accurately emphasized that Newtonian mechanics, while revealing order through the mathematical formulation of natural laws, could not explain the fundamental features of material order, particularly its stability. It was only with quantum theory that the emergence of stable elements or stable chemical compounds was understood. This is a central point of philosophical significance, and I was very pleased to see you place it at the heart of your presentation.

Of course, there is much to discuss regarding specific details. Let me begin with a minor critical remark: you may not have entirely escaped the danger (to which philosophers are often prone) of overestimating the capacity of language. For example, in addressing the question, "Are atoms real (*wirklich*)?" it is a mistake to assume that the word "real" (*wirklich*) has a clear and unequivocal meaning. The question should instead be framed as: "Is it a useful definition of the word 'real' to consider atoms as real?" Or, alternatively: "Is it more appropriate to describe atoms, or at least elementary particles, as ideas in the Platonic sense?" This question is structurally similar to another: "Is the double helix of the nucleic acid molecule alive?" Here, too, it is a matter

---

9. It is, most likely, a copy of his doctoral dissertation, "Atomic Order: An Introduction to the Philosophy of Microphysics," defended on June 18, 1966.

of defining terms, not of addressing an objective issue. The objective questions would be: "Does this molecule exhibit metabolism? Can it replicate and reproduce?" and so on.

One minor detail: on page 388, there is a typographical error concerning the weight of the moon. It should read 1025 instead of 1052.

Lastly, a didactic observation: I am not entirely certain whether it is advisable to present so much technical knowledge about atomic physics in the first part of the book. Scientists already know this material or at least know where to look it up, while non-scientists might find it daunting.

However, I am no expert in pedagogical matters, so perhaps you could consult a non-scientist among your friends on this point.

Should you come to Munich again, I would be glad to discuss further details of your book with you.

Best wishes for your continued work,
Yours sincerely,

H.

Letter 8. From Enrico Cantore to Werner Heisenberg
(August 23, 1967)

Fordham University, Department of Philosophy
Bronx, NY 10458

To His Excellency
Professor Werner Heisenberg, Director
Max Planck Institute for Physics and Astrophysics
8 Munich 23
Föhringer Ring 6
Germany

August 23, 1967

Dear Professor Heisenberg,

I am deeply indebted to you for your kind letter of the 16th of this month. During the many years of work my philosophical investigation has required, I encountered numerous difficulties and misunderstandings. Now, however, I know that it was all worthwhile. My joy is genuine, and my admiration for creative scientists, which has always been great, has grown even deeper. I am grateful to you for everything, especially for emphasizing the theme of natural order in the atomic domain. Unfortunately, the neopositivist perspective remains too dominant to allow for a full discussion of this topic in contemporary philosophy. Yet, as you rightly noted, it is indeed a central point of philosophical significance.

I will carefully follow your kind advice. Regarding the first part of my work, as you suggest, I will continue to inquire whether and to what extent it should be shortened. My reason for including such an extensive scientific introduction was methodological. It has always disconcerted me to observe how confidently philosophers express opinions about the nature and properties of the sciences without concretely considering the actual scientific data. As for the second part, I will review it again to avoid, as much as possible, the danger of "overestimating the capacity of language," as you aptly pointed out—particularly concerning the philosophical question of reality. The epistemological issues in this area are far too complex to permit a hasty resolution.

Finally, I will try to act on your primary recommendation: to publish my book as soon as possible. However, this brings me to one more request. Frankly, I would never have dared to ask you to read my manuscript if not for the hope that it might help bring my book to publication in a way that could contribute to addressing these problems. I have already submitted the manuscript for review to several individuals—a professor of physics, a professor of the history and philosophy of science, and a young physicist—and they have expressed approval for its publication. Nevertheless, I still fear, as I feared before, that it will not be easy for me to participate in the public scientific-philosophical dialogue. The prevailing mindset remains too strongly influenced by neopositivist prejudices, and I have no significant reputation to my name. It is therefore easy to anticipate that established publishers may resist publishing my work unless it is endorsed by a respected scientist and philosopher. That is why I have taken the courage to write to you. If you believe I can contribute something meaningful to this field, I kindly ask for your help.

I trust your experience and wisdom in knowing the best course of action in such cases. If I may offer a personal opinion—though I scarcely dare hope—I would like to request a foreword from you. Of course, this depends entirely on whether you think I have succeeded in achieving the goal of my investigation: to bring the philosophical structure of atomic physics to light. If this is the case, your remarks on the central significance of natural order in atomic physics would greatly contribute to the public recognition of this issue. Today, we are rightly concerned about the divide between the two cultures. It seems that the natural sciences have removed the central human motives of the unum, verum, bonum from the modern worldview. And yet, through the contemplation of order, these fundamental themes could acquire a renewed and deeper meaning. I think here of what you wrote in your lecture on Goethe and again, toward the end of your Geneva lecture (published in *On Modern Physics* under the title "Planck's Discovery and the Philosophical Problems of Modern Physics"), where, in reference to Goethe, you said: "In the beginning was the logos. To know this logos with respect to the fundamental structure of matter is the task of present-day atomic physics." [I have summarized your words.][10] These ideas could form the basis of an excellent foreword.

I will stop here. Thank you very much for your great kindness in allowing me to express this request. I trust you will act as your wisdom and

---

10. Heisenberg et al., *On Modern Physics*.

generosity recommend. For my part, I will strive to prove myself worthy of your trust by devoting myself with renewed courage to these problems. With best regards,

    Yours sincerely,

    (Enrico Cantore)

Letter 9. From Werner Heisenberg to Enrico Cantore
(September 1, 1967)

September 1, 1967

To His Reverend Excellency
Father Enrico Cantore
Fordham University
Fordham Road
Bronx, NY 10458 USA

Dear Mr. Cantore,

Thank you very much for your letter and the offprint, "Genetical Understanding of Science," which I look forward to studying. However, I would first like to address the questions raised in your letter. If I can assist you in quickly publishing your book, for instance, by writing a letter to a publisher or the like, I would be more than happy to do so. However, I must admit that I am not very familiar with the American publishing industry, so I am unsure which publisher I could recommend. My own book on elementary particles was published by Wiley, but that publisher is likely more interested in specialized scientific monographs than in philosophical works. You might consider reaching out to Harper & Row, which published my book Physics and Philosophy as part of the *World Perspectives* series. In that case, I would suggest corresponding with Dr. Ruth Nanda Anshen (112 East 61st Street, New York 28), who, based on my experience, might be interested in your philosophical book. You may also refer to me in your letter.

Unfortunately, I must decline your request at the end of your letter for me to write a foreword for your book. Over the past few years, I have received several such requests—for instance, from colleagues like Yourgrau and Jammer—and I have always declined. It seemed too difficult to say yes in one case and no in another. I hope you can understand my general position on this matter and excuse my regrettably negative response. That said, as mentioned, I would be glad to write a recommendation for a publisher.

With best regards,
Yours sincerely,

H.

Letter 10. From Enrico Cantore to Werner Heisenberg
(September 13, 1967)

Fordham University, Department of Philosophy
Bronx, NY 10458

To His Excellency
Professor Dr. Werner Heisenberg
Director—Max Planck Institute for Physics and Astrophysics
8 Munich 23—Föhringer Ring 6
Germany

September 13, 1967

Dear Professor Heisenberg,

Thank you very much for your kind letter of the 1st of this month. I fully understand your reasons for your general position regarding requested forewords, and I am grateful for your willingness to assist me through a letter to a publisher or something similar.

Here is the current situation: Recently, I sought advice and approached the director of the Fordham University Press. He immediately expressed his willingness to publish my book, provided that it genuinely appeals to scientists. Although he has published very few scientific works so far, apart from one on Wittgenstein, he is now eager to start. I therefore entrusted him with the manuscript for review. To be honest, this potential solution does not excite me, but it might be a good option. In this country, commercial publishers are solely interested in large print runs, so university presses seem to be the natural avenue for a work like mine. On the other hand, while Fordham University Press is relatively unknown among readers with scientific interests, the network of university presses has a comparatively good system for promoting new books.

If it seems appropriate to you, I would kindly ask you to write a letter to the publisher in a way that it could, if necessary, be photocopied and presented to more than one publisher. For this reason, I ask that the letter be addressed to me. (It might be better if it were written in English, but I can also translate it myself.) To provide some insight into the mindset of American readers, I am including a copy of a letter I recently received. The writer is a young physicist who has just completed his Ph.D. I sought his opinion as a representative of the general audience I hope to reach. (Please excuse the annotations I made to discuss the letter.)[11]

I will visit Dr. Anshen next Monday, as arranged by telephone. Thank you for this promising connection.

One more request: Should you plan to travel to the United States, I would like to know in advance. Fordham is earnestly striving to contribute to a synthesis between the two cultures.

If we had sufficient funds, we would gladly invite you to New York ourselves. Unfortunately, that is not possible, but we hope that you might consider including our university in a lecture tour.

In closing, allow me once again to express my gratitude and joy. It is deeply encouraging to experience that truly great people are also genuinely great in their humanity. May God bless you abundantly for it!

Yours sincerely,

(Enrico Cantore)

---

11. Cantore refers to some underlining marks in the original, which have not been reproduced here.

Attachment. Review from Bernard J. Burdick, PhD Physics (1970) from Case Western Reserve University

August 20, 1967
Cleveland, Ohio

Dear Fr. Cantore,

Finally I am able to report on my observations. I have the feeling that my analyses and reflections may be incomplete and not thoroughly comprehensible. This is in part due to the fact that I have not been able to reflect leisurely on the material; but mostly it is due to my unfamiliarity with philosophy of science in general. Hence, I am not able to place your work in its proper perspective with other similar studies. (This I hope to remedy in the near future.) Also, your work being only a preliminary introductory analysis of atomic order, I found myself left with a feeling of incompleteness and an only partially quenched (previously unknown) thirst. I think this is a good sign and one which speaks well for your book, for it indicates a resurgence of that feeling of awareness and awe.

Probably the most important thing to be learned, contrary to my previous experience with philosophy and with what philosophers usually say of science, is that philosophy can indeed comment significantly on the importance of science and at the same time keep the respect of the scientists.

Speaking frankly, I think the reason that I delayed so long in giving full attention to your work was an unconscious belief that I would be disappointed and this would be only a typical wishy-washy philosophic analysis of science. This was belied consciously by my personal acquaintance with you, but nevertheless persisted (being unconscious). Never, never could I have been more wrong! (I include these personal feelings to give you some appreciation of the psyche of a, I hope, typical graduate student and the effect your work might have.)

I think the greatest value of your work, and one which will make it appreciated and read by physicists, is the complete trust and respect you have in the achievements of science and in your own use of the scientific method for philosophic investigation (i.e., the inductive-genetic approach), and also in your accurate and authoritative use of scientific terminology.

It is encouraging to see a philosopher make use of physicist's ideas (at the same time, of course, being careful not to be misled by what they say when speaking philosophically but rather by observing their actions as "working physicists"); and then endowing their poorly worded ideas (philosophically speaking) with preciseness and clarity. It is also good to see that some present-day philosophers (such as yourself) are taking advantage of the lessons physics has learned and are not ignoring them—physics does offer to philosophy an excellent laboratory.

"Philosophy . . . particularly during recent centuries, has only too often been open to the accusation of being either a sterile restatement of the obvious or an arbitrary assertion of the unproved." (p. 625, fn 14)

Aside from the important philosophic conclusions, there is, for the practical physicist, another great contribution to be derived from philosophic reflections on science. This is the deeper meaning and greater significance that philosophy attaches to physical principles. This gives a physicist more of a "feel" for the behavior of matter. It is indeed enlightening to discover how a philosopher (an "outsider," one who should not—and you certainly do not—have preconceived views of physical reality) reflects on the phenomenon of atomic physics. This refreshingly different viewpoint can readily be appreciated by physicists who have not had sufficient time to become awed (i.e., graduate students).

It is really interesting to find out just to which principles a philosopher attaches importance—as these are often the ones overlooked and assumed by physicists. I especially enjoyed your explanation of the exclusion principle (p. 512, fn 10):

"In fact, were atoms not wholes, we could not find any clues at all to understand why two electrons were prevented from occupying the same energetic level." "Quantum numbers [are] manifestations of an intrinsic overall orderliness."

There can be no doubt that the greatest achievement of quantum physics is the discovery of an intrinsic order in mater. This is as it should be and is so self-evident that physics and physicists have taken it for granted! You have done a truly beautiful job of exposing this blandly assumed view as to what it really is—an awe-inspiring truth! There is no other fact of more importance to the foundations of physics (and upon which any science owes its existence) and to which so little attention is given. This is where the true beauty and great worth of philosophy shows through—much as the true beauty of language shows through in poetry.

The obvious advantage of philosophy over science (humanistically speaking) lies in its richer and more artistic interpretation of reality.

These truths are brought to life in the truly fine logical examination and brilliant clarification of the mechanistic conception of matter you present in depth. This exposition readily develops a truer appreciation for the achievements of quantum physics. Usually, physicists are aware of only a few basic experiments that are unexplainable by classical physics (viz., blackbody radiation, the photoelectric effect, specific heats, atomic spectra), but you have brought out the true philosophy inherent (and hence untenable) in classical, mechanistic physics. This is basically the reason for the lack of awe among physicists—the overall philosophic approach of mechanistic physics has never really been appreciated, in contradistinction to quantum philosophy.

Along these lines, I found your interpretation of quantization to involve quite an interesting and radically new view, namely, that quantization is the characteristic of autonomous totality—a beautiful idea!!

A word about the first volume and the division between the two volumes.

I found that you very accurately represented the achievements of science and covered quite completely all the necessary topics for use in later discussion in philosophy. I think you were wise to discuss these separately and not incorporate these together into one volume. The first volume was mostly a review for me, but it did clarify a few points and bring the whole subject into a proper focus.

As for the language, I think you have an excellent grasp of English and superb vocabulary. The sentence structure is usually very good, but sometimes you omit definite articles or use the wrong form of the word (specifically, adverbs for adjectives—e.g., relatively for relative). Only a minimum of proof-reading, however, should be needed.

Your expression of ideas is extremely precise and clear—all questions are stated unambiguously. Also the use of footnotes is excellent and constitutes one of the most important features of your book. (Don't let anyone talk you into putting these at the end of the book.)

There are a few things which I think might be covered in more detail, although maybe I am only looking forward to your next book.

The first is the role of mathematics in physics. In particular, the physical, predictions arising from purely mathematical analyses (e.g., the Dirac equations's prediction of positrons, sin of $1/2\hbar$ for electrons). Maybe this is further evidence of the intelligibility of matter. Along the same lines is the role of the conservation laws (momentum, mass-energy, charge, etc.). This is an extremely important idea in physics and a very useful tool for the physicist. Also, I notice you do not quote any of the recent physicists (Feynman, Sakurai, Gell-Mann). Feynman has a particularly comprehensive grasp of all of physics. Finally, would it be possible to accentuate your conclusions from the results and phenomena of other fields of physics? Specifically, the macroscopic order observed in superconductors and superfluids. This is a particularly fascinating field.

I'm sure I could go on for quite a while in praise of your work. In a typically American attitude, I would say that if you are selling stock in your book that I would be one of the first to buy some! I know your work will find wide acceptance because there is such a fantastic need for this type of study. Both of the "two cultures" will greatly benefit, as you have given physicists a finer appreciation for philosophy and you have shown philosophy the true path to follow in the examination of science and reality. May the "two cultures" henceforth share in the same wondrous and exciting *Weltanshauung* you have revealed!

I hope these ramblings of mine prove to be of some value to you and I hope that they, achieve their main purpose of encouraging you in your remarkable studies,

May I wish you all possible success in the publication of your work and in all your subsequent endeavors. If I can in any way be of assistance to you please do not hesitate to let me know. I will try to be more prompt next time.

As always,

(Bernard Burdick)[12]

Case-Western Reserve University
Department of Physics

Appendix.

12. Bernard Joseph Burdick graduated with a PhD degree in physics from the School of Graduate Studies at Case Western Reserve University on 28 January 1970. The title of his dissertation was "Single Particle Production in K-n Interactions at 5 BeV/c with a Visible λ or K^o." After his PhD he worked as a research scientist for MIT Lincoln Laboratory, Nichols Research Corp., and Torch Concepts. I thank the Case Western Reserve University Archives for helping me identify the author from the signature. His name is reported among the acknowledgments on the first pages of *Atomic Order*.

As regards the first part of your book, I have said very little. In order to give you a more meaningful analysis as you desire, I have looked it over again.

I am not so sure there is a dilemma involved since some of the readers will have a little scientific background and need only to be refreshed or to learn a few things they are not sure of (in particular, philosophers). This first volume will be a great aid to them as there are practically no good books on this subject at this level (i.e., discussion of the important. problems at a non-mathematical level). Whether such people would have an interest in your book is an open question. If they know a little science, they should develop a greater interest after having read the first volume and it seems as though you should be prepared for this in any case.

As for those people who have no scientific background—I don't think they will be even interested in the book. Their interest will be in direct proportion to the amount of exposure they have had to science above quantum mechanics.

As for those people who have had a good scientific background—herein lies your greatest logistical problem. These are the people who will be really interested in your work and who also will have little need or desire for the first volume. I rather think that they would be very hesitant to buy the first volume, and I doubt if they would do move than glance through it. I do think that it serves as an excellent verification of your credentials as a competent philosopher-scientist.

So you see that the value of your first volume is very dependent upon the qualifications of the reader, I'm not sure what I'd do but I don't think it would be feasible to shorten it nor to publish it as a separate book as I don't think it would be sellable. I also don't think it would be bought by more than 10–20 percent of the people interested if it were published as a separate volume. One possibility is to include them under one cover. This would produce a book of only slightly over four hundred pages (according to my calculations) and I doubt if this would have adverse effects in sales. I think the publisher would lose money if he tried to sell it as either as separate book or volume, though. Publication as a single volume, in my opinion, is probably the best solution although, admittedly, it is a compromise solution.

Letter 11. From Werner Heisenberg to Enrico Cantore (October 12, 1967)

October 12, 1967

To His Reverend Excellency
Father Enrico Cantore
Fordham University
Fordham Road
Bronx, NY 10458
USA

Dear Mr. Cantore.

Enclosed is the promised letter to the (as yet undetermined) publisher, which you may use at your discretion. I would be delighted if you achieve success either with Harper & Row or with Fordham University Press. It pleases me greatly to know that you have been so actively engaged at Fordham University in the philosophical interpretation of modern physics. In recent days, I have also discussed your efforts with Father Yanase, whom I met in Japan.[13] During our conversation, I learned about Patrick A. Heelan's book, *Quantum Mechanics and Objectivity*, which apparently also originates from your circle and has greatly interested me.[14] The next time I travel to America, I certainly hope to include a visit to Fordham University in my itinerary.

With warmest regards and best wishes,
Yours sincerely,

H.

---

13. Michael Yanase was a Jesuit-physicist, at that time member of the School of Natural Sciences at the Institute for Advanced Studies, Princeton.

14. See Heelan, *Quantum Mechanics and Objectivity*. Patrick A. Heelan (1926–2015) was an Irish Jesuit theoretical physicist and philosopher. Heelan had completed post-doctoral studies with Eugene Wigner (1960–1962) and, after a short period at Fordham, he moved to the Catholic University of Louvain for a second doctorate in philosophy under the supervision of Jean Ladriere. He was teaching at Fordham during 1965–1970, before moving at the State University of New York at Stony Brook, where he was head of the Department of Philosophy for twenty-two years. He was then professor of philosophy at Georgetown until 2013. Heelan maintained correspondence with Heisenberg from October 20, 1967 until November 8, 1974 (twenty-nine letters in the Archive of the Max Plank Society).

Letter 12. From Enrico Cantore to Werner Heisenberg (October 20, 1967)

Fordham University, Department of Philosophy
Bronx, NY 10458

To His Excellency
Professor Werner Heisenberg
Director, Max Planck Institute for Physics and Astrophysics
8 Munich 23
Föhringer Ring 6
Germany

October 20, 1967

Dear Professor Heisenberg,

A few days ago, I received your kind letter of the 12th and the enclosed recommendation letter for my book. You know that I have no words to express my gratitude. He who has granted you such great understanding and generosity will surely hear my prayer of thanks and bestow upon you the finest blessings.

I will keep you updated on the progress of my book. Many thanks also for the connection with Mrs. Anshen, which you facilitated. She is a wonderful woman, extraordinarily open-minded and helpful, and has opened up unforeseen possibilities for my work here.

In a few weeks, I hope to complete an article presenting my theoretical position on the synthesis between natural science and humanism. I will send the article to you. Thus, if you truly travel to the USA next spring, as Mrs. Anshen anticipates, your visit to Fordham could be highly significant for our budding initiative here. Thank you for including Fordham in your plans. Please, as soon as you make any decisions, let me know immediately. We would like to organize a public lecture by you and, if possible, seminars or meetings with professors and graduate students. I know the American mentality quite well and am convinced that your visit could do much to enhance the human understanding of natural science in this city, but it will require ample time to prepare everything thoroughly.

I recently had a look at P. Heelan's book. His method seems interesting but very different from mine. It is a joy for both of us that you show such interest in our writings. P. Heelan sends his kind regards. With heartfelt thanks,

Yours sincerely,

(Enrico Cantore)

Letter 13. From Werner Heisenberg to Patrick A. Heelan
(October 30, 1967)

October 30, 1967

Prof. W. Heisenberg

Dr. Patrick A. Heelan, SJ
Dept. of Philosophy
Fordham University
Bronx, NY 10458, USA

Dear Dr. Heelan,

Many thanks for sending me your book, which I first learned about through Dr. Yanase.[15] I was very interested to see that you, Dr. Büchel, and Dr. Cantore have undertaken such an intensive study of the philosophical situation in quantum physics.[16] I feel that I have found in the books of you and your colleagues a deeper understanding of the philosophy of quantum theory than in most of the current literature.

With best wishes to you and Dr. Cantore,
Yours sincerely,

H.

---

15. Heisenberg is referring to a letter Heelan sent him on October 20, 1967 (MPI, III/93/204/203–4). Heelan had sent Heisenberg a copy of *Quantum Mechanics and Objectivity*, his doctoral thesis on Heisenberg's thought, published in 1965.

16. Wolfgang Büchel (1920–1990) was a German Jesuit and philosopher who was particularly interested in the philosophy of physics. Büchel studied physics and philosophy and received his doctorate from the Ludwig-Maximilians-Universität in Munich in 1954 (*Die Relationalität des materiellen Seins*). He was a Jesuit and taught natural philosophy at the Jesuit College of Philosophy (Berchmanskolleg) in Pullach near Munich, where he had been an associate professor since 1956. From 1969 he was also professor of natural philosophy at the Ruhr University in Bochum, where he then moved permanently. At that time, Büchel had written *Philosophische Probleme der Physik* and *Wille, Wunder, Welt: Physikalisches Weltbild und christlicher Glaube*.

II.

"New York desperately needs scientific humanism."

The First Fordham Period and Heisenberg's Proposed Honorary Doctorate, December 1967–May 1968

CANTORE WENT SO FAR as to urge Fordham University to confer an honorary doctorate on Heisenberg on December 1, 1967. This effort highlighted Cantore's deep respect for Heisenberg's work and his desire to elevate the dialogue on science and philosophy within the university. Indeed, the purpose of the proposal went beyond mere academic formality; it was part of Cantore's project to make Fordham a center of scientific humanism, a place of active address, and perhaps even healing the split between the two cultures. But his enthusiasm is well apparent as he elaborates to Heisenberg on his plans, stressing the importance of the proposed visit to New York—not only for the honor it would bestow upon Heisenberg but also for the broader impact it could have on the intellectual climate at Fordham.

For his part, Heisenberg accepts with the grace and politeness of one who does not wish to refuse any honor, combined with a circumspect, half-expressed unwillingness due to health and the pull of all his other previous commitments. His reply shows an interest in helping Cantore's undertakings but is also qualified by practical limitations imposed by his situation. The following communication shows the delicate diplomacy that goes into making such an effort as Cantore tries to obtain the honorary degree and set a date for a visit that would suit Heisenberg's schedule.

However, events at Fordham take a dramatic turn. He is given the shocking news that he is no longer required at the university, despite all his efforts in developing scientific humanism. It comes when he is about to finalize plans for Heisenberg's visit, leaving Cantore disappointed and feeling like he does not know what to do. He expresses his vulnerability in those letters to Heisenberg, facing the implications of his dismissal and maybe the collapse of the whole program he has been working so hard to build.[1]

Therefore, it is important to cite Cantore's letter of February 23, 1968, for its emotional value. He frankly relayed his imminent departure from Fordham, expressing only his regret that this departure meant the Institute for Scientific Humanism was probably not to be. His desolation was increased by the suspicion that the university administration might never come to understand the import of his work or the promise of Heisenberg's participation. Cantore is exasperated by the "split between the two cultures" as he sulks that nobody understands him and helps him realize his mission.

Heisenberg responds encouragingly and realistically. His awareness of Cantore's challenges makes it possible to assure him of a potential reconsideration regarding a visit to Fordham if this could help the University in its decisions. Heisenberg is concerned with Cantore's plight because he knows only too well the high stakes of success or failure of the scientific humanism initiative. Further exchanges between Cantore, Heisenberg, and the authorities of Fordham University bring out the delicate balancing needed to keep the project's morale up. Cantore kept working for the program, sending letters to other influential academic community members to see if the initiative could be saved. Meanwhile, Heisenberg is guardedly participating, considering his path and what his involvement may mean.

The vulnerability and sincerity of Cantore's letters contrast sharply with the stereotypically impersonal and bureaucratic nature of the responses he received from the university. But amid all that frustration and

---

1. Cantore's vision for a center of scientific humanism at Fordham, while ambitious and noble, is thwarted by the complex interplay of administrative decisions, personal circumstances, and the broader cultural context of the time. The "Cantore Affair," as it might be called, can be read also from the side of Fordham University. See the appendix at the end of this book.

disappointment, there is a lingering sense of hope. After all, the vision and values of scientific humanism were only sidelined, but they are so powerful that they cannot be suppressed so easily.

Enrico Cantore is next to endure the vicissitudes of life within the American academy and workplace. Whereas the termination from Fordham was a heavy blow, Cantore stood resolute in his professional and academic determination to see through his lifelong vision of scientific humanism. Still, this new path forward was bound to become difficult. This section of the letter elaborates further on Cantore's personal and professional difficulties, revealing the weaknesses that make Cantore resilient and his unwavering support from Werner Heisenberg.

The letter Cantore wrote on March 24, 1968, presents a turning point in his series of communications with Heisenberg. As his position at Fordham was terminated suddenly and without any reason or explanation, Cantore started looking for other options that would allow him to continue his work on scientific humanism. He thinks of Rockefeller University, because of its interest in scientific research and the extensive investigation of the broader implications of science for human welfare. Cantore describes the ability to do significantly meaningful work within this new atmosphere—one that is interdisciplinary and welcomes inquiry into philosophy.

The letter is one of deep disappointment and frustration on Cantore's part, all tempered by a stubborn hope that his work will finally find a real home in some supportive academic environment. The very fact that he is appealing to Heisenberg for support suggests how serious Cantore realizes his situation to be. Without a compelling recommendation from someone of Heisenberg's prominence, his chances of a position at Rockefeller appear very slender. The fact that Cantore is so urgent and yet so humble in issuing the plea is poignant, for it reflects his recognition of the tenuousness of his situation.

The letter from Heisenberg, dated April 1, 1968, is compassionate but also quite practical. He considers both Cantore's difficulties and the general tension with which the humanistic tradition has regarded the concrete demands of modern science, as they transpire in his discussion and Cantore's experience in the academic world. Heisenberg's readiness to write a letter of recommendation to the president of Rockefeller University, Professor Detlev Wulf Bronk, indicated his continued support for Cantore despite his own work and troubled health.

Heisenberg's message to Bronk is that he appreciated Cantore's work and the need to replenish the gap between science and philosophy. Heisenberg underlines Cantore's peculiar capacity to unify such separate realms and expresses the value of his work for the philosophy of quantum theory and the hope that it can stimulate fruitful discussion at Rockefeller. This short letter can hardly be underestimated: abroad, the name of Heisenberg is widely known, and his prestige among scientific and academic circles is considerable.

The letters exchanged between Heisenberg and Cantore strengthen their feeling of brotherhood and solidarity. In this regard, Cantore appears thankful for Heisenberg's assistance in his letter dated April 7, 1968, where he tells Heisenberg that he is looking forward to his recommendation, which indicates open doors at Rockefeller. The letter from Cantore shows emotions; he is still hopeful and buoyed by what he seems to think is Heisenberg's kindness and generosity despite all odds.

The following months do not make Cantore's situation any more specific. His letter of May 16, 1968, says he has decided to return to Europe, at least for now, awaiting successive developments relative to his professional future. This letter is almost an epitaph, for Cantore has realized, as though at once, how many challenges lay ahead, but was ready to keep up with a life on scientific humanism, even though that such a road had to be followed within a very different context.

Heisenberg's closing letters in this chapter—dated April 2 and May 24, 1968—are shot through with his continued concern for Cantore and his determination to maintain their intellectual relationship despite the logistical and health difficulties both men face. Heisenberg's offer to Cantore to come once more to visit him in Munich again emphasizes the importance of their relationship and the value he places on their ongoing dialogue.

Letter 14. From Enrico Cantore to Werner Heisenberg
(December 1, 1967)

Fordham University, Department of Philosophy
Bronx, NY 10458

To His Excellency
Professor Werner Heisenberg
8 Munich 23
Föhringer Ring 6
Germany

December 1, 1967

Dear Professor Heisenberg,

I must inform you of something. I have nominated you for an honorary doctorate in philosophy here. You know why I have done this. It stems not only from a sense of gratitude that I owe to you but also from a profound conviction. Your contribution to philosophy, both as a physicist and as a philosopher, is almost immeasurable.

From a practical perspective, however, I would hardly have dared to take this step if I did not personally know of your willingness to help. For such a matter naturally entails an invitation to be present. I am well aware of how busy you are, and the distance is considerable. You have likely had to decline many such invitations, even from far more renowned universities than Fordham.

But I also know how committed you are to supporting the humanization of the scientific mentality. Your visit here would be very significant in this regard. New York desperately needs scientific humanism, and Fordham is trying to make a contribution in this area. If you could come here, it would be a great help to us.

I hope to send you a paper of mine on scientific humanism and the role of the university within the next two weeks. Although only theoretical, this paper might illustrate why I so deeply hope for your visit here. I want to make Fordham a center for dialogue between scientists and humanists.[2]

---

2. This is the first time in the correspondence with Heisenberg we find an explicit mention of this desire. And we learn that Heisenberg was asked to support this initiative.

The purpose of this letter is merely to inform you of the current situation. Around mid-December, the University Trustees will vote on the nominations for honorary degrees. Afterward, the university president would formally announce this to you. You should know that many people would be delighted if you could come. Dr. Anshen is particularly enthusiastic.

On the occasion of your visit here, you might consider taking a tour of the United States. The academic year-end ceremony, where honorary degrees are awarded, will take place on June 8. Naturally, universities and research centers would love to have you as their guest. Yesterday, I spoke by phone with Professor Weisskopf, who expressed interest in inviting you to Boston.[3]

Perhaps a plan of this kind could be considered. You could come here first, hold a few public lectures and seminars at Fordham, then spend a few weeks visiting the US, return here for the honorary degree ceremony, and finally depart. (The summer vacation at American universities begins almost everywhere in the second week of June.)

Fordham will, of course, take care of your travel to New York and all related arrangements.

If it is possible for you, please help us.

With heartfelt thanks, I wish you all the best.

Yours sincerely,

Enrico Cantore

---

3. Victor Weisskopf (1908–2002) was an Austrian-born American theoretical physicist. He did postdoctoral work with Werner Heisenberg, Erwin Schrödinger, Wolfgang Pauli, and Niels Bohr. During World War II he was group leader of the Theoretical Division of the Manhattan Project at Los Alamos, and he later campaigned against the proliferation of nuclear weapons. After World War II, Weisskopf joined the physics faculty at the Massachusetts Institute of Technology (MIT), ultimately becoming head of the department. He served as director-general of CERN from 1961 to 1966. Weisskopf was a member of the National Academy of Sciences, the American Philosophical Society, president of the American Physical Society (1960–1961), and the American Academy of Arts and Sciences (1976–1979). He was appointed member of the Pontifical Academy of Sciences in 1975.

Letter 15. From Enrico Cantore to Werner Heisenberg
(December 16, 1967)

Fordham University, Department of Philosophy
Bronx, NY 10458

To His Excellency
Professor Werner Heisenberg
Director, Max Planck Institute for Physics
8 Munich 23
Föhringer Ring 6
Germany

December 16, 1967

Dear Professor Heisenberg,

As you already know, we would very much like to have you as our guest here. Normally, at this time, the university president himself would have written to you to formally invite you to accept an honorary doctorate in philosophy. (The ceremony is scheduled for early June.) However, a slight delay has occurred this year. Two names on the list of candidates did not receive unanimous approval from the relevant committee.[4] Objections were raised against a federal judge and a very well-known woman.[5] As a result, the committee has postponed its decision on the entire list. It will likely take about another month before the formal invitations can be sent to the honorary degree recipients.

4. At the combined Boards of Fordham University decided on December 13, 1967, Mr. Tracy reported for the Honorary Degree Committee, and recommended the following nominees: Louis Armstrong, Dorothy Day, Elizabeth Sewell, Werner Heisenberg, Earl Warren, Paul Schweitzer. The Committee recommended also five alternatives: John Ronald Reuel Tolkien, Patricia R. Plante, Glenn T. Seaborg, John W. Gardner, McGeorge Bundy. The Committee report also mentioned the commitment to John Mary Lynch (1917–1999), known as Jack Lynch, prime minister (Taoiseach) of Ireland from 1966 to 1973 and 1977 to 1979, who was scheduled to be awarded the honorary degree in June 1968. McLaughlin also recommended the name of Fernando Lopez, at that time vice-president of the Philippines.

5. The "federal judge" in the list is most probably Earl Warren (chief justice of the US Supreme Court, Republican candidate for vice president of the USA in 1948), while the "very well-known woman" is probably Dorothy Day (founder of the Catholic Worker movement, active in the Socialist Party and close to the Communist Party, though never formally registered, converted to Catholicism in 1927).

I felt it my duty to inform you about this. I am also happy to take this opportunity to send you my warmest Christmas and New Year greetings. Hopefully, you will be able to respond to our great wish positively.

With great respect and gratitude,

(Enrico Cantore)

Letter 16. From Werner Heisenberg to Enrico Cantore (December 20, 1967)

December 20, 1967

To His Reverend Excellency
Father Enrico Cantore
Fordham University
Fordham Road
Bronx, NY 10458, USA

Dear Mr. Cantore,

Thank you very much for your letters informing me about the planned honorary doctorate from Fordham University. I gratefully regard such an honor as a sign of friendship and shared interest in the central problems of contemporary philosophy. I am delighted that a number of colleagues at Fordham University evidently share these interests.

For this reason, I would gladly visit Fordham University at some point, though I would prefer not to make firm promises regarding the timing. I have become somewhat less inclined to travel, as it is tiring for me, and I need to select a time when I am not too burdened by other commitments. I trust you will understand this.

Once again, my heartfelt thanks to you and your colleagues for your kind intentions, and my best wishes for the Christmas season.

Yours sincerely,

H.

Letter 17. From Leo McLaughlin, Fordham University, to Werner Heisenberg (February 1, 1968)[6]

Fordham University, Office of the President
Bronx, NY 10458

February 1, 1968

Dr. Werner Heisenberg
Director
Max-Planck-Institute for Physics and Astrophysics
University of Munich
6 Föhringer Ring
Munich, Germany

Dear Dr. Heisenberg:

I am pleased to inform you that the Trustees of Fordham University have empowered me to extend to you a cordial invitation to accept the honorary degree of Doctor of Science.

Honorary Degrees will be presented at the annual University Commencement at 10:30 a.m., Saturday, June 8, 1968, on our Rose Hill Campus, Bronx, New York.

You will be interested in knowing that a Committee consisting of Faculty, Administration, Alumni and Students recommended you for this honor. The Board of Trustees then approved the recommendation.

Please let me know if you will be willing to accept this honor which is the highest Fordham University can bestow.

Sincerely yours

Leo McLaughlin, SJ[7]

---

6. The official correspondence with Fordham University concerning the conferment of the honorary doctorate to Werner Heisenberg is found here in Letter 17 (February 2, 1968), Letter 20 (February 12, 1968), Letter 23 (February 23, 1968), Letter 24 (February 29, 1968), Letter 30 (April 2, 1968).

7. Leo McLaughlin (1912–1996) was president of Fordham University during 1965–1968. In an obituary article from *The New York Times* (August 18, 1996) we read: "As a transforming president of Fordham from 1965 to 1969, he opened up the curriculum beyond traditional theological courses, encouraged academic experimentation, fought for higher faculty salaries even as he was turning the established

Letter 18. From Enrico Cantore to Werner Heisenberg
(February 6, 1968)

Fordham University, Department of Philosophy
Bronx, NY 10458

To His Excellency
Professor Dr. Werner Heisenberg
Max Planck Institute for Physics and Astrophysics
8 Munich 23
Föhringer Ring 6
Germany

February 6, 1968

Dear Professor Heisenberg,

Thank you very much for your kind letter of December 20. We are delighted to hear of your interest in the planned honorary doctorate from Fordham University. You truly have many friends and admirers at this university. Even students in the humanities are enthusiastic about the idea that you might visit us one day. This is also why I proposed your name for a doctorate in the humanities. (The official title, according to state regulations, is Doctor of Humane Letters.)

Jesuit faculty inside out, brought Marshall McLuhan to Fordham as a professor and even wrested the university from Jesuit control. . . . At a time of intellectual and social ferment, Dr. McLaughlin established his credentials as a man of the times, not only devoting a full morning a week to meeting with students and listening to their ideas but spending a good part of the rest of the week working to put many of them, plus a few of his own, into effect. 'New ideas,' he once said. 'That's all that counts in today's world, new ideas—and so few people have them.' Dr. McLaughlin's overriding idea was to transform Fordham from a somewhat sleepy and hidebound religious college into a vibrant academic colossus. . . . Within two years, Dr. McLaughlin was being credited with bringing about some of the most significant and controversial changes in the college's history. But none were as momentous as the one he engineered before leaving the presidency to become chancellor at the start of 1969." Thomas, "Leo McLaughlin, Jesuit Teacher," 51. McLaughlin had to resign from the presidency at the end of 1968. A recent essay on the history of Fordham University states: "In an unprecedented development, in December 1968 the Roman authorities of the Society of Jesus forced the resignation of the president of Fordham University, Father Leo McLaughlin, SJ, because of fiscal mismanagement and replaced him with Father Michael Walsh, SJ, the former president of Boston College." Shelley, *Fordham*, 219–20. For a detailed account on McLaughlin's presidency and role at Fordham, see Shelley, *Fordham*, 395–410.

Today, however, to my great disappointment, I learned that you were contacted regarding the awarding of a title in the natural sciences. I immediately sought clarification, and it had to be admitted that a bureaucratic misunderstanding had occurred. The university president was clearly in agreement with my proposal, but the secretary drafting the letter primarily knew you as a scientist—and the president signed the letter without noticing the discrepancy. Please accept our apologies. I will soon rectify this oversight.

Your visit here would be extraordinarily significant for scientific humanism. Americans greatly need this humanism and are strongly influenced by the natural sciences. Thus, if a creative scientist and renowned philosopher like yourself were to publicly advocate for this humanism in New York, your efforts would be particularly fruitful. You could also greatly help us make our contribution in this area more effective.

The day of the honorary degree ceremony coincides with the so-called "University Commencement," which is scheduled for June 8. Perhaps you could come here before the exam period begins on May 13. You could deliver a few public lectures, visit the various institutes that would be delighted to host you as a guest, and then return here for the commencement. Alternatively, if such an extended stay would be too demanding, you could come after the exams conclude at the end of May, give a few lectures, and receive the title—all within the same week. Unfortunately, the honorary doctorate cannot be awarded *in absentia*.

We fully recognize that Fordham cannot properly honor you. But we also know that you will do whatever is possible to assist us, and for that, we thank you most sincerely. Please provide me with a definitive answer so that I can make the necessary arrangements. Naturally, we will cover all your expenses.

With kind regards and the hope that you can agree to our request, I remain,
Yours sincerely,

Enrico Cantore

Letter 19. From Werner Heisenberg to Enrico Cantore (February 9, 1968)

February 9, 1968

To His Reverend Excellency
Father Enrico Cantore
Fordham University
Fordham Road
Bronx, NY 10458, USA

Dear Mr. Cantore,

Thank you very much for your letter. In the meantime, I have received the official notification from the president of your university that I have been nominated for an honorary doctorate from Fordham University. It does not seem particularly important whether the honorary doctorate is in the natural sciences or the humanities—though I greatly appreciate your thoughtful gesture in emphasizing the humanities in this context. I have not yet responded to the president's letter because the suggestion of coming to Fordham University on June 8 poses significant difficulties for me. I was ill for several weeks and need to take care of myself, but I already have many speaking commitments scheduled. In May, I am supposed to travel to Bologna for two weeks, followed by meetings in Bonn at the end of May and sessions in Berlin on June 11 and 12. A longer stay in America during this period is therefore out of the question, and I must honestly say that even a brief trip would be too exhausting for me. I understand that an honorary doctorate cannot be awarded *in absentia*, but I would like to inquire whether it might be possible to postpone the ceremony to a later date, at which time I could include a visit to Fordham University as part of a longer trip to America. Could you let me know whether this would be possible (perhaps even delaying the conferral of the degree by a year)? I will respond to the president of your university only after hearing back from you.

Once again, my sincerest thanks for all your efforts.

With best regards,
Yours sincerely,

Signed: W. Heisenberg

Letter 20. From Fordham University to Werner Heisenberg
(February 12, 1968)

Fordham University, Board of Lay Trustees
Bronx, NY 10458

Dr. Werner Heisenberg
Director
Max Planck Institute for Physics and Astrophysics
University of Munich
Föhringer Ring 6
Munich, Germany

February 12, 1968

Dear Dr. Heisenberg:

The President of the University recently conveyed to you the news that the Trustees had voted to offer you an honorary doctorate at our Commencement exercises in June.

Father McLaughlin has conferred with Father Enrico Cantore, who is certainly more aware than anyone else at the University of your distinguished scholarly background, and it is the judgment of both of them that, if you wish, it might be more appropriate to have conferred upon you the degree of Doctor of Humane Letters rather than that of Doctor of Science.

The University would, of course, be guided by your preference.

I should appreciate it if you would let Father McLaughlin know of your decision in the matter.

Very truly yours.

Michael J. Sheahan

cc:
Rev. Leo McLaughlin, SJ
Rev. Enrico Cantore, SJ

Letter 21. From Enrico Cantore to Werner Heisenberg (February 17, 1968)

Fordham University, Department of Philosophy
Bronx, NY 10458

February 17, 1968

Dear Professor Heisenberg,

Thank you very much for your letter of the 9th of this month. Your kind understanding and your willingness to support our efforts in scientific humanism are very encouraging.

I fully understand your reasons for postponing a visit to America. I have done everything I could to have the conferral of the honorary doctorate delayed. Unfortunately, the policies here are inflexible in this regard. The honorary doctorates must be conferred exclusively during the Commencement, and they are unwilling to make commitments for the next academic year. As Professor Weisskopf remarked yesterday during my visit to MIT, the Fordham administration does not truly understand the situation. It is not Fordham that honors you, but rather you who honor Fordham by accepting the degree.

I feel deeply torn. On one hand, I do not want to impose anything unreasonable upon you. On the other hand, I fear that if you cannot attend the Commencement, we will miss a great opportunity for our program. The Commencement provides a wonderful setting to launch our initiative. (I am enclosing a copy of the program for your reference.)

Allow me to add just a few points. As Dr. Weisskopf also noted, it is not certain that a longer stay in America would be less tiring for you than a shorter one. After all, it might be difficult for you to fend off the attention of your many admirers.

I will conclude here. I am confident that if it is at all possible for you, you will do everything in your power to help us—despite the lack of understanding from some individuals here. (Such a lack could perhaps be seen as concrete evidence of the divide between the two cultures. All our administrators here belong to the humanities!) I would like to revisit my earlier proposal. Monday and Tuesday, June 3 and 4: two public lectures for academics and university students in New York. Then, a few days of rest. (You could, for example, travel to Princeton, as

Dr. Weisskopf suggested, and stay with friends there, or stay with Dr. Anshen—she is very enthusiastic about the possibility, as she informed me by telephone—or even plan a longer stay in America.)

I am ready to do everything possible to ensure your comfort. Perhaps you already know that you could fly out in the morning from here and arrive in London by evening, instead of spending the entire night on the plane. This would significantly reduce fatigue.

If this plan is too demanding for you, please be assured that I fully understand, and my gratitude will remain as great as ever.

Please let me know your decision. With respect, Yours sincerely,

Enrico Cantore

Attachment. The program mentioned in the text

SCIENTIFIC HUMANISM AT FORDHAM UNIVERSITY

A Practical Program for Overcoming
the Split Between the Two Cultures

—A Memorandum—

Part I—General Considerations

I. Theoretical Foundations

1. The Humanistic Crisis of Modern Man. Modern man is in a state of humanistic crisis. He is profoundly puzzled and dissatisfied. One main reason for this situation is the emergence of great, unexpected problems due to scientific-technological advances. The enormously complex achievements of science threaten to make the world more of a prison than a home for man. On the other hand, the new knowledge and power offer unsuspected opportunities for the improvement of man himself. The humanistic crisis of modern man is that of finding a self-identity in the scientific-technological world.

2. Nature and Need of Scientific Humanism. The crisis can be met only through the creation of a new humanism. This should help man find his identity and ideal in the context of reality taken as a whole. The new humanism should integrate the scientific perspectives of the present with the inalienable humanistic values of the past and thus make scientific progress fully human. To this end, two essential steps are necessary:

(a) study in depth the implications of science affecting man as a whole;

(b) rethink systematically the humanistic values in the light of the present scientific perspectives.

3. The Humanistic Responsibility of the University. The university should be the privileged place for developing the new humanism, for the university is essentially a center of intellectual research aimed at forming the whole man. These pages intend to present a humanistic practical program now being prepared by Fordham University.

II. Practical Guidelines

Preliminary: The Justification of a New Program. The universities are increasingly recognizing their humanistic responsibility. They frequently organize interdisciplinary seminars and symposia and make the proceedings available to the public. Yet the results cannot, as a rule, be considered satisfactory enough. The reason is probably to be found in the fragmentary and superficial nature of the method adopted. A new program, therefore, is justified if it can provide a systematic approach. The practical lines of such an approach can be as follows.

1. *Provide a Permanent Forum for a Dialogue on the Humanistic Meaning of Science.* The creative scientist is the privileged witness to the humanistic values of science. In general, however, he is not professionally prepared to make his views fully acceptable to the humanistic community. The humanist, for his part, is usually not in a position to realize for himself the profoundly humanistic nature of science. The so-called two cultures are the outcome of such estrangement. Hence the first practical step for a systematic approach to scientific humanism: establish a permanent forum of dialogue. The creative scientist, reflecting on his personal experience, will present concrete instances of science as a humanistic attitude. The interested humanist, sympathetic to the new problematic, will contribute to the dialogue by asking pertinent questions. This first step is most significant for developing mutual understanding and cooperation.

2. *Pursue Systematic Research into the Humanistic Problems of Science.* Once the problems have been realized in their complexity and importance, the need for systematic academic research will become apparent. Such problems, in fact, are much too complex to be solved through mere discussions. The second practical step, therefore, will consist in the establishment of a school of research. In it, students with scientific background should examine concrete issues in detail under the guidance of both experienced scientists and humanists. The aim of their research should be to integrate the values of science into a profoundly rethought humanistic framework.

3. *Contribute to the Improvement of Education.* Theoretical research has no real significance unless it helps man to better himself. This can ultimately be achieved only by improving education. As a consequence, the third step in the systematic attempt to solve the problems of scientific humanism can be carried out in two phases. The first phase will consist in making the results of the dialogue between scientists and humanists immediately available to educated public opinion. This can be achieved through a series of books containing the proceedings of the encounters. These books will gradually introduce the public to the humanistic nature of science, hence progressively remove the misunderstanding between

the two cultures. The second phase will consist in making available to the public the results of the profound academic research and translating these results into college textbooks. Then the progress of science will finally be fully significant for man as a whole.

Part II—The Contributions of Fordham University to Scientific Humanism

I. The Contributions of The Past

Fordham's active interest in scientific humanism has already a tradition. Here are the principal initiatives sponsored in the past.

1. An Investigation of the Thought of Teilhard de Chardin. Fall 1963: Fr. Maurits Huybens, SJ, European editor of the *International Philosophical Quarterly*, delivered a series of six open lectures as an introduction to the thought of Teilhard. The audience, averaging six hundred per lecture, included students and faculty from neighboring colleges and universities. The discussions of Teilhard's thought were continued by the Fordham faculty throughout the school year by means of periodical seminar meetings. This led to the foundation of a specialized Research Institute, named after Teilhard.

Summer 1964: the Institute organized a five-week Workshop of Study and criticism of Teilhard's writings. The Workshop was concluded by a one-week Conference (August 17–21), held at the Fordham Campus. The papers delivered on that occasion were published under the title *Proceedings of the Teilhard Conference 1964*. Both the book (1,100 copies sold so far) and taped copies of the Conference papers have found a good reception by the public.

2. The Fordham Interscience Conference on "The Images of Man." A theme of central interest to both scientists and humanists is currently that of evolution and the emerging images of man. The Teilhard Institute organized a week-long conference on this theme at the Onchiota Conference Center, Sterling Forest, NY (June 21–25, 1965). The group comprised about fifteen scholars drawn from a number of universities and a wide range of fields (physics, chemistry, biology, the anthropological and social sciences, philosophy and theology). The undertaking was made possible through a grant of the Ford Foundation.

3. Lecture Series on "The Sacred and the Secular." To study the implications of the so-called "secular Christianity," the Institute organized a series of open lectures spread over the school year 1965/1966. Among the prominent speakers were John Smith of Yale, Joseph Sittler of Chicago and Harvey Cox, the author of the Secular City. The lectures were well attended and received.

II. PRESENT CONTRIBUTIONS

The encouraging experience of the past has prompted Fordham to review and expand its contributions to scientific humanism. The emphasis during this school year (1967/1968) has been on reorganization and initiation of a systematic dialogue between scientists and humanists.

1. *Organizational Activities*. One faculty member with a specialized background in both science and philosophy, and a considerable experience of the interdisciplinary dialogue, was hired by the Graduate School of Arts and Sciences with the main task of studying the concrete possibilities in this field. He has been consulting widely both with the various Fordham departments and with professors of other academic institutions in the Northeastern area of the country. The results of his endeavors find their expression in some current initiatives and a planned program for scientific humanism.

2. *Some Current Initiatives*.

(a) *Interdisciplinary Seminar on Structure and Order*. Members of eight Fordham Departments (Biology, chemistry, physics, mathematics, philosophy, English, sociology and psychology) convene regularly to discuss the typical way of studying the intelligibility of reality proper to each of the various specialties involved;

(b) *Seminars on the Philosophical Commitments in the Origins of Modern Science*. Dr. A. C. Crombie, Head of the History of Science Department at the University of Oxford and well-known author, has been invited to give four seminars on this theme to the Fordham faculty and to guests from neighboring universities (April 1–10, 1968). Dr. Crombie will also give an open lecture to the student body on the theme: Galileo, Science and Philosophy;

(c) *Nomination of Professor Werner Heisenberg to an Honorary Doctor's Degree in the Humane Letters (to be conferred at the '68 Commencement)*. With this nomination Fordham intends to honor an outstanding creative physicist for his exceptionally important contributions to scientific humanism,

(d) *Courses in Scientific Humanism.* Two new courses have been made available during this school year for all college students who wish to overcome the gap between the two cultures,

(e) *Hiring of Specialized Faculty.* Up to the present the Fordham faculty includes already two members with doctor's degrees both in physics and philosophy. A very prized addition will be Dr. Satose Watanabe, an internationally known theoretical physicist and philosopher of science, who is scheduled to begin his work here during the coming school year.

Part III—Fordham's Planned Program in Scientific Humanism

Fordham's planned contributions to scientific humanism can be outlined under the following headings.

1. *Galileo Encounters.* Consistently with the fundamental need of a mutual understanding between the two cultures, the first practical service that Fordham can provide is the establishment of a permanent forum of encounter between scientists and humanists. The aim is the systematic exploration of the meaning of science for modern man. The method is dialogical. Creative scientists, drawing from their own personal experience and reflection, are invited to illumine the main humanistic implications of their science in a series of public talks. Each talk will be followed by a panel discussion. Panels are to be comprised of scientists and humanists drawn both from inside and outside the Fordham faculty. The name of Galileo stands for the spirit of openness and sympathetic understanding which should animate the encounters. The plans are to explore in depth one science each year. Each yearly series of encounters should be terminated by a roundtable. Physics is scheduled to be the first one to be taken up.

The encounters are meant to lead to a greater public understanding of the nature of science and its relationship to daily living. Hence they will be free of charge and open to the public, in particular to the students and the faculties of the neighboring colleges and universities. Since nothing of this kind exists up to now in the Greater New York area, Fordham could play a leading role in this field and provide a much-desired service to the academic community of the area.

2. *Volume Series on the Humanistic Meaning of Modern Science.* The proceedings of the Galileo Encounters are intended to be published in book form, one volume for each of the sciences explored. It is thus

hoped that a whole series can be developed dealing with the humanistic meaning of modern science. What would justify the appearance of such a series is the fact that the currently available literature usually does not explore systematically and in depth the meaning of the various sciences. The series could then be of considerable service to the educated public, especially college students and graduates. In particular, the series could provide a primary source of information for curriculum specialists concerned with the need of making science understandable to liberal arts students.

3. Interdisciplinary Seminars. A third initiative planned at Fordham is the creation of a number of standing seminars on subjects of common interest to science and humanities (e.g., creativity in science and the humanities, the images of man according to the developing sciences, investigation of integrative education, scientific humanism and man's practical responsibilities, etc.). The seminars are to be open to all faculty members of Fordham and of neighboring institutions actively interested in their subjects. The establishment of such seminars would meet a much-felt need in the area of the Greater New York. Fordham, as a Christian university, seems to be in a special position for succeeding in this type of undertaking. A Christian university, in fact, should be fundamentally concerned with man as a whole. Hence Fordham could provide a unique service to the academic community desirous of discussing how the human person can find again its central place in the development of culture.

4. Public Information Service on Scientific Humanism. There is a great need for timely and reliable information about publications and initiatives bearing on scientific humanism. A complementary activity to the foregoing ones would be that of providing such information. Participants in the Galileo encounters and the interdisciplinary seminars, as well as all other interested people, could easily contribute the information about new books, articles and events they happen to be acquainted with. The information would then be made available to the public by means of a newsletters. The newsletters, in turn, could be significant in providing a focusing point for scientific-humanistic discussions and could serve to test the desirability of a specialized periodical on scientific humanism.

5. Graduate Research Project in Scientific Humanism. As obvious from the nature of the subject matter, activities of the type listed can have no more than an introductory, though indispensable, character. The problems of scientific humanism, in fact, can only be solved by a deep investigation carried out by specialized full-time researchers. As a consequence, the long-range service that Fordham can give in this area can also be envisaged as including the establishment of a center for graduate research in scientific humanism.

The researchers should be young scientific graduates willing to investigate, historically and philosophically, concrete problems arising out of science. Foundation fellowships could be obtained to support the students. The involvement of the faculty would not add much to their normal duties. As is done in other universities (Oxford, Chicago, etc.) this involvement would consist mainly in advising and guiding research, each faculty member according to his specialty.

Letter 22. From Enrico Cantore to Werner Heisenberg (February 23, 1968)

Fordham University, Department of Philosophy
Bronx, NY 10458

His Excellency
Professor Dr. Werner Heisenberg
Director, Max Planck Institute for Physics
8 Munich 23
Föhringer Ring 6
Germany

February 23, 1968

Dear Professor Heisenberg:

I have an important news to communicate to you at once. When a few days ago I sent you a copy of my memorandum about the project on scientific humanism at Fordham, it seemed that the adoption of such a project here was practically certain. I had interviewed all chairmen, and the majority of them was clearly favorable. Also experts from other universities and institutions (Rockefeller University, the Polytechnic Institute of Brooklyn, The New School of Social Research, the National Science Foundation) were enthusiastic in supporting the idea. A meeting to decide the issue had already been summoned for the 29th of this month. Then, all of a sudden, I was informed that Fordham had no interest in my presence here any more, so I should leave here at the end of the school year.

You may imagine how much disappointment there was, in me and in those who supported the initiative. The meeting of the 29th has been cancelled. I do not know the reasons for such a sudden change—although I suspect that a certain influence of the two cultures is not absent.[8]

I had to inform you at once about this development, because you have been so kind, understanding and helpful that I owe you my best consideration and respect. Now it would be rather embarrassing for you

---

8. Cantore's somewhat cryptic statement here alludes to the opposition he faced from certain scientists at Fordham. To fully grasp its implications, it should be read in conjunction with the appendix.

to come here expecting that Fordham be moving in this field, when probably nothing will happen. I am really sorry about this whole event, but I cannot do anything to change it.

I plan to leave here at the beginning of May, as soon as classes will be over. A very encouraging aspect concerning this undertaking has been the reaction of the physicists I have interviewed so far with the hope of inviting them here during the next school year for the Galileo Encounters. So far I had the adhesion of Drs. Weisskopf and Cyril Smith of MIT,[9] Dr. Uhlenbeck of Rockefeller,[10] Dr. Seeger of the National Science Foundation.[11] Dr. Rabi[12] has promised to grant me an interview soon. Seeing that such great men like you and these scientists are so willing to cooperate on a project of scientific humanism gives me courage, even in the midst of the disappointment. I hope that something can be done elsewhere. I shall ask Dr. Rabi whether something could be done at Columbia.

---

9. Cyril Stanley Smith (1903–1992) was a British metallurgist and historian of science. He is most famous for his work on the Manhattan Project where he was responsible for the production of fissionable metals. A graduate of the University of Birmingham and MIT, Smith worked for many years as a research metallurgist at the American Brass Company. In 1961, he moved to MIT as an Institute professor with appointments in both the Departments of Humanities and Metallurgy. He applied the techniques of metallurgy to the study of the production methods used to create artefacts such as samurai swords. Among his notable works is the essay *Sources for the History of the Science of Steel*.

10. George Eugene Uhlenbeck (1900–1988) was a Dutch-American theoretical physicist. He was professor of physics (1960–1974) and emeritus professor of physics (1974–1988) at Rockefeller University. Other institutional affiliations included University of Michigan and University of Utrecht. His research interests included nuclear physics, random processes, statistical mechanics, and cosmic rays. Member, National Academy of Sciences (1955); president, American Physical Society (1959); awarded National Medal of Science (1976); awarded Wolf Foundation Prize in Physics (1979).

11. Raymond John Seeger (1906–1992) was an American physicist. From 1952 until 1970, he worked at the National Science Foundation where he became deputy assistant director, then retired as a senior staff research associate. He taught at the American University from 1954 until 1972.

12. Isidor Isaac Rabi (1898–1988) was an American physicist who won the Nobel Prize in Physics in 1944 for his discovery of nuclear magnetic resonance, which is used in magnetic resonance imaging. He was also one of the first scientists in the United States to work on the cavity magnetron, which is used in microwave radar and microwave ovens. When Columbia created the rank of university professor in 1964, Rabi was the first to receive that position. A special chair was named after him in 1985. He retired from teaching in 1967, but remained active in the department and held the title of university professor emeritus and special lecturer until his death. His works include Rabi, *My Life* and Rabi, *Science*.

I am really sorry to have inconvenienced you. But I still hope to be able to meet you in the States during the next school year. Still a word concerning my book. Harper & Row are phasing out their division dealing with philosophy of science. Bobbs-Merrill have examined the book, but said that they have already too many commitments extending over two or three years. I shall now follow the advice of Dr. Weisskopf and try at MIT Press.[13] Dr. Weisskopf is also quite favorable to the book itself.

Please have the most sincere expressions of thank and esteem from me. I know how much I owe you and I cannot finish marveling at your great kindness and understanding.

May the Lord reward you abundantly
Sincerely yours,

Enrico Cantore

P. S. You will excuse me for my writing in English. I have realized that I make too many mistakes when typing in German. But, please, write in German to me: I have no difficulty at all in understanding it.

---

13. The MIT Press was formally established in 1962, just a few years earlier. It began operating as an independent publishing entity following its separation from John Wiley & Sons. Carroll Bowen was appointed as the Press's first director.

Letter 23. From Werner Heisenberg to Leo McLaughlin (February 23, 1968)

February 23, 1968

Professor Dr. W. Heisenberg

The President
Fordham University
Professor Leo McLaughlin
Bronx, NY 10458, USA

Dear Sir:

May I express my sincere thanks for the invitation of Fordham University to accept an honorary doctorate at the annual University Commencement on Saturday, June 8, of this year. I consider this invitation as a great honor, especially since I am closely connected with some of the professors of your university by common interests in the philosophy of science. It is also on account of this common interest that—as I understood from the letter of the executive secretary of your university—Father Enrico Cantore had the very kind idea to confer upon me the degree of a Doctor of Humane Letters in order to emphasize the close connection of the philosophy of modern science with very old philosophical problems which have been discussed both in ancient Greece and medieval times. I am very glad to follow this suggestion of Father Enrico Cantore.

It is obvious that I will be able to accept this great honor only if I can be present at Fordham University on June 8th, and this is the only question for which I hesitate to give a definite answer. During the last two month my health had not been too good, and I will only after the Easter holidays be able to see to what extent I can stand traveling and lecturing. Therefore I would be grateful if I could postpone my definite answer to the time after the Easter holidays.

May I ask you to convey my thanks also to your colleagues in Fordham University,

Sincerely yours

H.

Letter 24. From Leo McLaughlin to Werner Heisenberg (February 29, 1968)

Fordham University, Office of the President
Bronx, NY 10458

Professor Dr. W. Heisenberg
Max-Planck-Institut fur
Physik und Astrophysik
8 München 23
Föhringer Ring 6
West Germany

February 29, 1968

Dear Professor Heisenberg:

Thank you so much for your letter of February 23, 1968.
 Of course, you may postpone your definite answer until after the Easter holidays. I hope to hear from you on or about April 24, 1968.
 I hope that you will be well enough to travel and lecture.

Sincerely yours

Leo McLaughlin, SJ

Letter 25. From Wolfgang Büchel to Werner Heisenberg (March 8, 1968)

Philosophische Hochschule Berchmanskolleg Pullach
8023 Pullach bei München

March 8, 1968

Dear Prof. Heisenberg:

I am enclosing a carbon copy of the letter I wrote today to P. Cantore, as per your request.

Allow me to once again express my heartfelt thanks for your kind visit, and for now, I remain with my best regards,

Yours sincerely,

W. Büchel

***

Attachment: a letter to Fr. Cantore

Pullach: March 8, 1968

Dear Father Cantore,

Yesterday, Professor Heisenberg kindly visited us here at Berchmanskolleg—on the occasion of the lecture I gave during the academic celebration of the Feast of St. Thomas.[14] During his visit, he showed me your letter, in which you informed him about your removal from Fordham and the failure of your plans to establish a "Center for Scientific Humanities [sic]." At the same time, he mentioned that he had been invited by Fordham University to receive the degree of *Doctor honoris causa*. However, after receiving your letter, he is undecided about whether to travel to Fordham or not, as he had primarily accepted the invitation with the intention of supporting your plans. He then asked me to write to you—privately and unofficially—as follows:

If Prof. Heisenberg can, by attending, ensure that your removal is reversed or that the founding of the "Center" is definitively secured, he will gladly make the effort to travel to Fordham. However, if his visit cannot prevent your removal or advance the establishment of the "Center," he wishes to postpone the trip to Fordham for now because of his fragile health—he was ill for a long time during the winter.

May I therefore kindly ask you to inform me—privately and unofficially—whether Professor Heisenberg's presence could secure the founding of the "Center" or reverse your removal. I will then forward your response to Professor Heisenberg.

Sending you, and also P. Heelan, warm greetings on behalf of myself and Professor Heisenberg,

Sincerely.

---

14. On March 7, 1968, at the University's annual academic feast in honor of Thomas Aquinas, Büchel lectured on matter and mind in the presence of Werner Heisenberg, drawing heavily on the latest findings in physics research.

Letter 26. From Enrico Cantore to Werner Heisenberg (March 24, 1968)

Fordham University
Bronx, NY 10458

Professor Dr. Werner Heisenberg
Director, Max Planck Institute for Physics
8 Munich 23
Föhringer Ring 6

March 24, 1968

Dear Professor Heisenberg:

I hope you have already received my answer to your message forwarded to me by Fr. Büchel. I really do not know how I can thank you for so much goodness and generosity. Unfortunately, however, nothing can be done here at Fordham concerning my plans dealing with scientific humanism.

Everything happened so suddenly. The situation appeared develop in an encouraging way. The Vice President for Academic Affairs had already even called a meeting of all departments heads, together with professors of other institutions, to decide the foundation of the Institute. He was strongly in favor of the initiative. Also the majority of the participants was favorable. Then someone from the Administration ruled that I had to leave Fordham at the end of the school year. The meeting announced never took place.[15]

The reasons for such a negative and sudden decision are not clear to me. I was told that there were financial difficulties.[16]

15. The administration was headed at that time by Dr. Paul Jacob Reiss (born 1930), a young professor of sociology, both executive vice president for Lincoln Center and dean of the Liberal Arts College at Lincoln Center. Reiss earned a BS, *magna cum laude*, from the College of the Holy Cross, an MA in sociology from Fordham University in 1954, and a PhD from Harvard University. Reiss taught at Marquette University (1957–1963) and then at Fordham (1963–1985). Reiss was recruited by Father Joseph Fitzpatrick from Marquette University. Father McLaughlin's successor as president of Fordham, Father Michael P. Walsh, promoted Reiss to academic vice president of the university while retaining his position at the College of Liberal Arts at Lincoln Center, where Cantore was doing research at the Department of Philosophy.

16. It is possible that the initiative to establish an additional institute in such an innovative field as the dialogue between philosophers and scientists initially met with McLaughlin's favor—especially given that the University had the intention to award

This event has been the hardest blow of my whole life. Hence your generous offer of help has been all the more significant to me.

Now, after having consulted numerous people, I may tell you that it is still possible for me to work successfully for scientific humanism in New York. However, this will not be possible without some further assistance from you. Thus I dare to disturb you again.

Various persons, notably Professor John Wheeler of Princeton with whom I was speaking two days ago, suggest that I try to get a post at the Rockefeller University.[17] The reasons are as follows. First, the specific purpose of the university. Differently from other universities, this institution is wholly dedicated to the furtherance of science and its consequences for the human welfare. In view of this aim, the present President, Professor Bronk, who is very able and broad-minded, has added philosophy to the various fields of interest of the university. The second reason is the fact that Rockefeller is not organized in rigid departments, as universities here usually are, but consists of numerous groups of researchers. Cooperation among the various groups is possible and desired. The third reason is the method of work, host of the activity there consists in research and seminars, rather than in formal classes. Finally, the type of students. They are outstanding young people, with a good background in science, willing to do original research.

This is the way I see the possibility of bringing some significant contributions of mine to Rockefeller. First, by means of original research and writing. I have now matured many ideas which I could easily put

---

Heisenberg an honorary doctorate. However, this came at a time of significant financial distress at Fordham, for which McLaughlin, as president, bore responsibility: "In March 1968, he said that 'we have reached a moment of truth' and revealed that Fordham would have to raise a minimum of $3.5 million by June 30, 1969" (Shelley, *Fordham*, 408). The initiative may have suffered a halt with McLaughlin's impending resignation as president imposed by the Jesuit General Curia in Rome. See Shelley, *Fordham*, 387–410.

17. John Archibald Wheeler (1911–2008) was an American theoretical physicist. He was largely responsible for reviving interest in general relativity in the United States after World War II. Wheeler earned his doctorate, age twenty-one, at Johns Hopkins University under the supervision of Karl Herzfeld, and studied under Breit and Bohr on a National Research Council fellowship. For most of his career, Wheeler was a professor of physics at Princeton University, which he joined in 1938, remaining there until 1976. At Princeton he supervised forty-six PhD students, more than any other professor in the Princeton physics department. Wheeler left Princeton University in 1976 at the age of sixty-five. He was appointed as the director of the Center for Theoretical Physics at the University of Texas at Austin in 1976 and remained in the position until 1986, when he retired and became a professor emeritus.

into writing, if I could be in a suitable place with free time, possibility of discussion with specialists, and a good library. Second, by fostering a systematic dialogue with the creative scientists. Some friends inside Rockefeller tell me that my presence there would be helpful in this regard, since at present there seems to be not enough dialogue between scientists and philosophers. The reason appears to be the fact that the philosophers there belong mostly to the analytic or Neo-positivistic mentality. At any rate, I know from experience that a dialogue with the scientists of Rockefeller is possible and fruitful. In the past, I have already had occasion of meeting Drs. Pais[18] and Uhlenbeck, the physicists; and Dr. Dobzhansky,[19] the geneticist. Their reactions were quite encouraging. Another reason for my desire to be at Rockefeller is the possibility of having bright students, with scientific background, willing to do research into the humanistic aspects of science.

In sum, it seems to me that Rockefeller would be an ideal place for developing the program I have in mind. For the moment, of course, I would not plan to develop anything like an Institute for Scientific Humanism. But, in practice, this University would give me the possibility of carrying out many of the activities that I intended to do in the planned Institute here. As for the future, all possibilities would remain open. Being in New York, in the neighborhood of very great universities interested in the dialogue like Columbia, it is likely that the opportunities for scientific humanism would increase with time.

I feel very much embarrassed to disturb you again, but I think your help is indispensable. Would you, please, be so kind as to write to the President of the Rockefeller University and recommend me for a post there? Even if you do not know him personally, this does not matter. The prestige of your name suffices. I was told that there is no hope of being accepted there except upon recommendation of outstanding scientists.

---

18. Abraham Pais (1918–2000) was a Dutch-American physicist and science historian. Pais earned his PhD from University of Utrecht just prior to a Nazi ban on Jewish participation in Dutch universities during World War II. He was a physics professor at Rockefeller University until his retirement. In the late 1970s Pais became interested in documenting the history of modern physics.

19. Theodosius Grigorievich Dobzhansky (1900–1975) was a prominent geneticist and evolutionary biologist. He was a central figure in the field of evolutionary biology for his work in shaping the modern synthesis. Born in Ukraine, Dobzhansky emigrated to the United States in 1927, aged twenty-seven. From 1962 until his retirement in 1971 he was professor at Rockefeller University.

In addition, can I beg you to write to the President soon? My Superiors allow me to continue to stay in the US if I can succeed in finding a post before the end of this school year.

I would never have thought that working for scientific humanism would present so many difficulties. I feel strongly tempted to abandon the effort. So I owe you an immense gratitude for your understanding and willingness to help. If I shall succeed in bringing some important contribution to this field, this will certainly be due to your help. So may God reward you abundantly for everything,

Very thankfully yours

(Enrico Cantore)

P.S. I enclose here some sheets from an official publication which can give you an idea about the structure of Rockefeller U. The address of the President is:

Professor Detlev Wulf Bronk[20]
President, The Rockefeller University
New York 10021
USA

[Attachment: an excerpt from the 1966–1967 Catalogue, Rockefeller University, five pages which describe Functions and Organization, Graduate Studies and Research at RU.]

---

20. Detlev Wulf Bronk (1897–1975) was a prominent American scientist, educator, and administrator. He is credited with establishing biophysics as a recognized discipline. Bronk served as president of Johns Hopkins University from 1949 to 1953 and as president of The Rockefeller University from 1953 to 1968. Bronk also held the presidency of the National Academy of Sciences between 1950 and 1962.

Letter 27. From Werner Heisenberg to Enrico Cantore
(April 1, 1968)

April 1, 1968

Prof. W. Heisenberg

To His Reverend Excellency
Father Enrico Cantore
Fordham University
Fordham Road
Bronx, NY 10458, USA

Dear Mr. Cantore.

I was very sorry to hear about the great disappointment you experienced at Fordham University. I recently became aware of the significant differences of opinion within the circles close to you regarding the relationship between the humanistic tradition and modern natural science, particularly through the discussion following a lecture by Father Büchel. This makes it all the more important not to shy away from any effort to initiate or sustain dialogue between both sides. Therefore, I urge you not to be too discouraged.

As per your request, I have written to Professor Bronk, and I would be very pleased if you were given the opportunity to continue your work there.

With best wishes,
Yours sincerely,

H.

[Attachment: a copy of the letter of recommendation to Prof. Bronk, Rockefeller University][21]

---

21. See Letter 28 (April 1, 1968).

Letter 28. From Werner Heisenberg to Detlev Wulf Bronk, Rockefeller University (April 1, 1968)

April 1, 1968

Professor Detlev Wulf Bronk
President, The Rockefeller University
New York 10021 (USA)

Dear Professor Bronk:

Professor Enrico Cantore has given me a note that he is intending to apply for a post at the Rockefeller University, and he has asked me to send you a few lines of recommendation.

I know Dr. Enrico Cantore through several conversations on the philosophical foundations of quantum theory and through his book on *Atomic Order—An Introduction to the Philosophy of Microphysics*. I had the impression that Dr. Cantore is one of the few people who can help to bridge the gap between modern science and the old humanistic philosophical tradition of Europe. His book is certainly a very valuable contribution in this direction. Therefore I hope that he will be given the opportunity to pursue his line of thinking at a place which seems suitable to him for this purpose.

Yours very sincerely.

H.

Letter 29. From Werner Heisenberg to Wolfgang Büchel (April 2, 1968)

April 2, 1968

Dear Mr. Büchel,

Thank you very much for your efforts and for forwarding Father Cantore's letter to me along with the copies. I am sorry to hear of Cantore's great disappointment at Fordham University. At his request, I have now written to the President of Rockefeller University, where Cantore intends to apply.

I was very pleased with your recent lecture in Pullach, but the discussions there also showed how difficult it is to build a bridge between the two schools of thought, theology and philosophy on the one hand, and modern natural science on the other. I was also very interested in your review of Heelan's book. It seems to me that the most important thing at the moment is that these problems are being seriously thought about at all; the fact that they are being thought about is more important than the particular result the thinker comes to.

Best wishes

H.

Letter 30. From Werner Heisenberg to Fordham University (April 2, 1968)

April 2, 1968

The President
Fordham University
Bronx, NY 10458, USA

Dear President:

Many thanks for your letter. Unfortunately it has turned out that I will not be able to come to the States during this summer term, since I have too many other obligations in Europe. Therefore I regret very much that I will not be able to accept the honorary degree which you so kindly have offered, and I have to apologize that I have on account of the other obligations to come to this negative answer.

Yours very sincerely,

H.[22]

---

22. With this letter Heisenberg seems to decline the honorary doctorate. His name was already not present in the list of Fordham's honorary doctorate recipients in the Board of Trustees' Meeting dated March 13, 1968 ("Resolved that the following honorary degrees may be conferred upon the following named persons: Doctor of Laws: the most reverend Terence J. Cooke; Doctor of Letters: Elizabeth Sewell; Doctor of Letters: Paul Schweitzer." The Board of Trustees' meeting reports that "Three honorary degrees have been authorized for the June 8, 1968 Commencement. It has been understood that a fourth candidate might be proposed"). Heisenberg's name does not appear as nominee, neither here nor in the Board of Trustees' Meeting dated January 19, 1969. The 1969 candidates will be: "Hon. Edward M. Kennedy, Dr. Arthur Burns and Mr. John Olin (Honorary Doctor of Laws, LLD); Mr. Irving Berlin (Doctor of Humane Letters, LHD)."

Letter 31. From Enrico Cantore to Werner Heisenberg
(April 7, 1968)

Fordham University
Bronx, NY 10458

April 7, 1968

Dear Professor Heisenberg,

Just a brief word to express my deepest gratitude for your encouraging letter and the recommendation letter of the 1st of this month. I pray that the Lord Himself may richly reward you for it.

I have just written to Professor Bronk and will keep you informed about the outcome.

Best Easter greetings.
Yours sincerely,

(Enrico Cantore)

Letter 32 From Enrico Cantore to Werner Heisenberg
(May 16, 1968)

Fordham University, Department of Philosophy
Bronx, NY 10458

To His Excellency
Professor Dr. Werner Heisenberg
Max Planck Institute for Physics
8 Munich 23
Föhringer Ring 6
Germany

May 16, 1968

Very esteemed Professor Heisenberg:

In a few days I shall leave New York for Europe. For the moment no definite opening has presented itself concerning my future, so it is perhaps better that I wait in Europe for the development of events.

I would certainly be very glad if I could thank you in person for your very great kindness and generosity in helping. I shall be in Aschaffenburg (Institut St. Marla, Obernauerstrasse 40) from about May 25 to about June 11. I could make a trip to Munich sometime during the last week of May or, even better, after my definitive departure from Aschaffenburg.

If this would not be too much of a disturbance for you, and in any way it could be possible to agree on a common date, I would certainly feel it a duty and a joy for me to come to see you again.

At least, please have the most sincere expressions of my deeply felt obligation. I want to assure you, too, that I really intend to do my best for contributing to the scientific-humanistic dialogue also in the future.

Gratefully yours,

(Enrico Cantore)

Letter 33. From Werner Heisenberg to Enrico Cantore (May 24, 1968)

May 24, 1968

To His Reverend Excellency
Father Enrico Cantore
Institut St. Maria
875 Aschaffenburg
Obernauer Strasse 40

Dear Mr. Cantore,

Thank you very much for your letter. I would be delighted to meet with you in Munich, but I am unsure whether a suitable time can be arranged, as I have many commitments this summer. Perhaps, after your arrival in Aschaffenburg, you could call my office in Munich to inquire about possible times for our meeting.

    With best regards,

    H.

III.

"The philosopher needs the benevolent help of the creative scientist."

Second Meeting and the Writing of Heisenberg's "Physics and Beyond," July 1968–June 1969

CANTORE WROTE HEISENBERG FROM Heythrop College in England, where he had temporarily taken up residence after leaving Fordham University. The Jesuit is there as a man on a mission: he is committed to taking on the most significant scientific minds of the age. Most directly, Cantore is obsessed with the meaning that modern science has for the philosophical conception of the human person that the Catholic church promoted after the Second Vatican Council. Cantore is determined to explain this latter commitment, given the depth and detail of his preoccupations with discussions that have yet to be discussed.

"Principles for Scientific Humanism" is the name Cantore gave to this comprehensive program, which he outlined in his letters to Heisenberg and represents an essential stage in the intellectual quest that was his life's journey. The program involved the systematic elaboration of a series of books dealing with the relationship between science and humanistic values as a way to address the two-culture problem. It is this ambition of Cantore to build a dialogue to bridge this gap, bold and innovative, that reflects his deep-seated desire to further the creation of a more unified whole in human knowledge.

In his replies, Heisenberg offers a blend of practical support and philosophical guidance. He recommended that Cantore visit Munich in October 1968, when the two could relax and talk about the philosophical underpinnings of science, thereby proving how important he was according to these interactions and engagements. The fact that Heisenberg was eager to invest his time and vision for Cantore shows that both shared respect and intellectual fraternity.

By February 20, 1969, Cantore could communicate some good news to Heisenberg: "After extensive consultations, I am glad to report that the general agreement (including the personal wish of the General) is that I should continue to develop the work begun. Thus, I am now getting ready for going to New York again. I expect to be there by the end of March."

Yet this was a difficult period. Cantore had increasingly serious problems finding an institutional base in New York City, and these grew even stronger as he tried to remain permanently in the country. Although Heisenberg worked hard to obtain positions for him at, e.g., Rockefeller University, Cantore never had a warm reception there, especially from academics of the dominant analytic-philosophical school of thought who opposed his approach in all possible ways.

However, the exchange between Cantore and Heisenberg showed pressure. In the latter's letters, there was a mixture of hope and frustration when Cantore tried to address institutional obstacles threatening to drive his efforts down the drain. For his part, Heisenberg remains a firm friend who gives moral encouragement with practical support. Even more indicative of the ambivalence of their relationship are the negotiations that continued until his last days with Harper & Row for the translation of Heisenberg's *Physics and Beyond*, which Cantore had been eager to translate based on Heisenberg's suggestion. Despite his deep understanding of Heisenberg's philosophical ideas, Cantore ultimately stepped aside when Harper & Row decided to use a different translator, A. Pomerans, recognizing the practicalities and sensitivities involved in such a decision.[1] The publisher's decision "saddened" Heisenberg (June 6, 1969).

---

1. A glaring delay in the translation—hastily done and, according to Cantore, improvable for a second edition—caused the publication of *Physics and Beyond* to be postponed until 1971, while the German version, *Der Teil und das Ganze*, was promptly released by Piper in 1969.

This chapter sheds more light on the central figure of Ruth Nanda Anshen, the editor of the *World Perspectives* series.² The two sets of correspondence—first between Heisenberg and Anshen, and then between Cantore and Anshen—reflect the honor Anshen felt in working with Heisenberg and her careful evaluation of Cantore's abilities as a translator. These relationships reflect broader issues in academic publishing at the time, when intellectual and commercial interests were often closely intertwined.

Despite facing professional challenges, Cantore remained fully committed to the larger cause of scientific humanism. His determination is clear in his plans to organize private meetings with professors in New York, where he hoped to spark discussion and build momentum for his ideas, even without formal institutional support. This energy and persistence highlight just how important this project was to him.

Regarding Rockefeller University's rejection of Cantore's application, Heisenberg wrote: "I suspect, as you do, that these are related to the conflicts between various philosophical directions, and for now, it seems the logicalist direction is dominant," but he also added: "I hope you will not lose heart in what you have set out to do, and if I can help you in any way, I will gladly do so" (June 6, 1969).

---

2. The correspondence with Anshen is included in Letters 40, 43, 46–47, 50, 52, 54, 60–64.

Letter 34. From Enrico Cantore to Werner Heisenberg (July 4, 1968)

Heythrop College
Chipping Norton, Oxon (England)

July 4, 1968

Very esteemed Professor Heisenberg:

I am writing to you these few lines first of all to inform you about my present address. According to the decision of my Superiors, I shall spend my time mainly here doing research and writing in the philosophy of science. This place is suited being in the vicinity of Oxford.

Above all I desire to express my great gratitude to you for your understanding and support. I cannot express in words what I feel inwardly. Now, in particular, I am so thankful for your invitation of coming to your Institute for a leisurely discussion of the philosophical structure of science. I am deeply convinced that the philosopher needs the benevolent help of the creative scientist to really understand science, the concrete science. You have been so good to offer me this marvelous opportunity. Of course, there are moments when I feel hesitant in accepting this generous invitation. This is due to my realizing my immense ignorance. Yet, in this connection too, I know that you are so kind and understanding, hence I feel confident again.

Now that I can give my permanent address, I am sure that your secretary will remember to notify me the date of the coming of Professor von Weizsäcker,[3] as soon as this date has been decided upon. By the way, I was thinking that possibly I could introduce myself to the same Professor by sending him a copy of my paper on Scientific Humanism.[4] To this end I would appreciate if I could have his address.

---

3. Carl Friederich von Weizsäcker (1912–2007) was a German physicist, astrophysicist, and philosopher. He was the longest-serving member of the team of scientists who did nuclear research in Nazi Germany during World War II, under Werner Heisenberg. After the war he resumed his scientific studies, devoting himself to teaching and working with the Max Planck Institute. As early as 1944 he had re-proposed the nebular hypothesis for the formation of the solar system. He was professor of philosophy at the University of Hamburg. In 1989 he was awarded the Templeton Prize for his scientific contributions.

4. Cantore is probably referring to the manuscript attached to Letter 37 (August 13, 1968), later published as Cantore, "Scientific Humanism and the University."

Once more, my most sincere thanks to you and my respectful greetings.

Truly yours,

Enrico Cantore

Letter 35. From Annemarie Giese to Enrico Cantore
(July 26, 1968)

July 26, 1968

Secretary

To His Reverend Excellency
Father Dr. Enrico Cantore
Heythrop College
Chipping Norton, Oxon (England)

Dear Reverend,

A few days ago I had the opportunity to ask Professor v. Weizsäcker about his plans for his next stay in Munich. I would therefore like to inform you, as agreed, that Professor v. Weizsäcker intends to be in our Institute in Munich during the week of October 7 to 12. I must add, however, that Prof. v. Weizsäcker has not yet been able to finalize his schedule for this fall and that postponements are therefore possible.

Since you want to send your work to Prof. v. Weizsäcker in preparation for a meeting, I will be happy to give you his permanent address: Prof. Dr. C. F. Frhr. v. Weizsäcker, Philosophical Seminar of the University of Hamburg, 2 Hamburg 13, Von-Melle-Park 6.[5]

With best regards

A. Giese

---

5. The document mentioned has been sent, and is recorded here as Letter 37 (August 13, 1968).

Letter 36. From Enrico Cantore to Werner Heisenberg
(August 10, 1968)

Heythrop College
Chipping Norton, Oxon (England)

August 10, 1968

Dear Professor Heisenberg:

Your very kind and quite efficient secretary, Miss Annemarie Giese, has already Informed me about the probable time Professor von Weizsäcker will be at your Institute during the next October. I wish to thank her for this piece of information.

Now I am writing in order to define somewhat more precisely the work that I can carry out during my next stay in Munich. I outline here what I consider a possible program. At the same time, I request your kind comments and suggestions.

My central reason for a trip and stay in Munich is the marvelous opportunity you have suggested of discussing the profound implications of science, especially atomic physics, for modern man. You know how much I have this topic at heart. I am really very thankful to you for offering me this occasion of examining these themes in depth. I am firmly convinced that only by learning from creative scientists can the philosopher come to realize the humanistic significance of science. Since you are the creator of atomic physics, with all its deep-reaching philosophical implications, I am overjoyed of this opportunity.

Certainly I do not intend to take too much of your time, but for a leisurely study and reflection, which allow me the possibility of asking meaningful questions, I think that I should spend several weeks in your city, coming to see you at appointed times only.

Another topic I would gladly submit to you would be a general plan of attack of these philosophic-humanistic problems raised by science. I am now developing a systematic program for the years to come, both of research for myself and fixture students of mine. I would be most thankful if I could hear your opinion about it.

Finally, when being in Germany, I would greatly like also to become acquainted with the German contribution to the philosophical reflection on science.

My question now is to know whether you think this plan of work reasonable, and what is likely to be the best time to carry it out. Personally I am pretty much free to come whenever the circumstances are best suited to the aim. If October is the right month, very well. Otherwise I am ready to come later on. As for the length, it seems to me that probably one-month stay would be reasonable.

Best, heartfelt thanks for all your kindness. I feel very embarrassed in taking so much of your precious time. You may be sure that my thankfulness and respect are very great indeed.

Sincerely Yours,

(Enrico Cantore)

Letter 37. From Enrico Cantore to Werner Heisenberg
(August 13, 1968)[6]

PRINCIPLES FOR SCIENTIFIC HUMANISM

General Introduction. The Problem of Man in the Age of Science: The Humanistic Challenge to Philosophy

VOLUME I—AN EXPERIENTIAL DISCUSSION OF KNOWLEDGE FOUNDATIONS FOR A SELF-UNDERSTANDING OF MAN

Part I—The Observable Structure of Knowledge

Intr. The Structure of Knowledge as Accessible to Observation

1. The Psychogenetical Dimension of Knowledge

2. The Sociological Dimension of Knowledge

3. The Historical Dimension of Knowledge

Concl. The Epistemological Problem of Man (his individuality, his changeability)

Part II—The Interiorizable Structure of Knowledge

Intr. The Structure of Knowledge as Accessible to Interiorizing Reflection (the self-discovery of man through reflection)

1. The Structure of Perception:

   A) Awareness of objects

   B) Awareness of types (*Gestaltwahrnehmung*)

2. The Structure of Intellection:

   A) The Activity of the Knower (Abstraction, model-making, etc.)

---

6. The documents presented here do not belong to any conversation and were sent by Enrico Cantore, likely for discussions with Carl Friedrich von Weizsäcker and Werner Heisenberg between October 7 and October 31, 1968. Attachments: "Scientific Humanism and the University" (manuscript, twenty-seven typewritten pages), later published as "Scientific Humanism and the University"; "Appendix to Scientific Humanism and the Role of the University" (manuscript, thirteen typewritten pages); "Principles for Scientific Humanism" (document outlining a series of five books on the topic). We include a transcription of this document because it traces ideas that are further developed in subsequent works: *Scientific Man* (1977), *Science and Human Dignity* (unpublished, 1977), and *Christ-Wisdom and Science* (unpublished, 2006).

B) The Receptivity of the Knower (Evidence, certainty, etc.)

3. Knowledge as a Response of the Whole Person.

4. The Dynamism of Knowledge.

Concl. Foundations for a Humanistic Self-Understanding of Man

Volume II—An Experiential Discussion of Reality
The Self-Understanding of Man in the World

Intr. Experience, the Key to the Self-Understanding of Man in the World

Part I—*The Observable Structure of Reality*

Intr. The Structure of Reality as Accessible to Observation

1. The Formal Dimension of Reality (mathematico-logical features of reality)

2. The Interactive Dimension of Reality (the physico-chemical-biological features of reality)

3. The Self-Determining Dimension of Reality (the psychoranthropological features of reality)

Part II—*The Interiorizable Structure of Reality*

Intr. The Structure of Reality as Accessible to Interiorizing Reflection

1. Reality as Intelligibility: The Discovery of Pervasive Truth

2. Reality as Harmony: The Discovery of Pervasive Beauty

3. Reality as Positivity: The Discovery of Pervasive Goodness

Concl. Connaturality of Man with Reality

Volume III—An Experiential Discussion of Creativity
The Self-Understanding of Man for the World

Intr. Experience, the Key to the Self-Understanding of Man for the World

Part I—*The Observable Structure of Creativity*

Intr. The Structure of Creativity as Accessible to Observation

1. Intellectual Creativity: Man, the Maker of New Intelligibility

2. Artistic Creativity: Man, the Maker of New Harmony

3. Practical Creativity: Man, the Maker of a New World

Concl. Man's Humanization of Reality

Part II—The Interiorizable Structure Of Creativity

Intr. Structure of Creativity as Accessible to Interiorizing Reflection

1. Creativity as Attraction: The Dynamism of Value

2. Creativity as Engagement: The Discovery of Freedom

3. Creativity as Tension: The Struggle for the Ideal

Concl. The Drama of Man the Creator

Part III—Interpersonal Dimension of Creativity:
The Humanistic Significance of Religion

Intr. The Ineffable Presence: An Experiential Analysis

1. Creativity as Calling: The Personal Discovery of God

2. Creativity as Response: The Self-Discovery of Man

3. Creativity as Partnership: Communion toward Fulfillment

GENERAL CONCLUSION

THE CONTINUING DIALOGUE: Wisdom as the Ideal and the Mission of Man

Letter 38. From Werner Heisenberg to Enrico Cantore (August 13, 1968)

August 13, 1968

To His Reverend Excellency
Father Enrico Cantore
Heythrop College
Chipping Norton/Oxon
(England)

Dear Mr. Cantore,

Thank you very much for your letter. I fully agree with your plans and would suggest that you stay in Munich in October. Until the end of September (with a few interruptions), I will likely be on vacation at my country house. In October, however, I will be in Munich most of the time, with only a few short commitments in Bonn toward the end of the month. Since Professor von Weizsäcker will also be in Munich for some time in the first half of October, you can take the opportunity to have discussions with him.

With best regards,

signed Heisenberg

(On behalf of Prof. Heisenberg, who departed after this letter was dictated, sent by A. Giese)

Letter 39. From Enrico Cantore to Werner Heisenberg (September 12, 1968)

Heythrop College
Chipping Norton, Oxon (England)

September 12, 1968

Dear Professor Heisenberg:

This short note is only to inform you about the date of my coming to Munich. Because of several reasons I shall have to delay my arrival until the 6th of October. On the 7th I shall contact your secretary and, if possible, come to see you.

I want to express once again my most sincere thanks for this unique opportunity you are giving me to understand from within the humanistic meaning of scientific creativity, and its philosophical significance.

Thankfully yours

(Enrico Cantore)

Letter 40 from Werner Heisenberg to Ruth Nanda Anshen (November 4, 1968)

November 4, 1968

Mrs. Dr. Ruth Nanda Anshen
14 East 81st Street
New York 28, NY, USA

Dear Mrs. Anshen,

Thank you very much for your letter and for the invitation to New York. I cannot yet say whether my trip to America next spring will happen. For now, I would prefer not to think about new travels.

My planned book is making good progress and has now grown to about 150 typewritten pages. Perhaps I will send you this first part in the not-too-distant future, but first, I would like to write a foreword to make the overall plan clearer.

In recent weeks, Father Cantore, whom I once recommended to you for his book *Atomic Order*, visited me. I also showed him the manuscript of my planned book, and he expressed interest in translating it into English. Of course, I have not discussed anything definite with him yet, but I would like to hear your opinion on whether you consider him suitable as a translator. He has lived in America for a long time, so he believes he has a full command of the English language (likely in its American version).

With many warm regards, also to Richard Courant,[7]

Yours sincerely,

H.

---

7. Richard Courant (1888–1972) was a German American mathematician. He is best known by the general public for the book Courant and Robbins, *What Is Mathematics?* His research focused on the areas of real analysis, mathematical physics, the calculus of variations and partial differential equations. He wrote textbooks widely used by generations of students of physics and mathematics.

Letter 41. From Enrico Cantore to Werner Heisenberg (November 7, 1968)

Heythrop College
Chipping Norton, Oxon (England)

November 7, 1968

Dear Professor Heisenberg:

It is now a week since I left your Institute. With the passing of time I realize more and more the importance of my sojourn there for my whole work. Knowing your modesty, and the pressure of your occupations, I have delayed writing until the present. However, since this is a duty for me, please allow me to express my thankfulness in a few, simple words.

My sojourn in Munich was a unique occasion for getting acquainted with science as a concretely lived activity. The result has been that my esteem for the scientists has increased immensely, and my determination to serve the cause of scientific humanism has been vastly strengthened.

Knowing how much I owe you, I cannot say more than a heartfelt Thank you! But my intention is to prove, through facts, the seriousness of my determination to serve the scientific community.

As for you who embody so well the ideal of the scientist-humanist, my wish and prayer is that you may be able to develop right to the end the marvelous program to which you have been called and to which you have so generously dedicated all your life,

Sincerely yours,

(Enrico Cantore)

P.S. Still a word of admiration and thanks for your outstanding Secretary, Miss Giese.

Letter 42. From Annemarie Giese to Norbert Bischof
(November 18, 1968)[8]

November 18, 1968

Dr. Norbert Bischof
Max Planck Institute for
Behavioral Physiology
8131 Seewiesen über Starnberg

Dear Dr. Bischof,[9]

On behalf of Father Dr. Cantore, I am returning the book *Productive Thinking*, which you kindly lent him, with our sincere thanks.[10]

With best regards,

A.
Secretariat of Prof. Heisenberg

---

8. Cantore was probably returning a book he had borrowed from Norbert Bischof during his stay in Munich.

9. Norbert Bischof (born 1930) is a German behavioral psychologist and system theorist. He worked with Konrad Lorenz at the Max Plank Institute for Behavioral Psychology in Seewiesen (1966–1975).

10. See Wertheimer, *Productive Thinking*. Max Wertheimer's book is considered a landmark in creativity research, exploring various creative thought processes, from solving geometric problems to understanding the development of Einstein's theory of relativity. The book is a significant contribution to the psychology of thinking, particularly in Gestalt psychology, and distinguishes between insightful ("productive") and automatic ("unreflected") thinking. Wertheimer's direct conversations with Einstein offer unique insights into the latter's thought process. The book's enduring relevance is highlighted by its applicability to modern fields such as psychology, education, neuroscience, and philosophy.

Letter 43. From Ruth Nanda Anshen to Werner Heisenberg (November 21, 1968)

World Perspectives—Harper & Row Publishers
14 East 81st Street
New York 28, NY

November 21, 1968

Dear Professor Heisenberg:

What a great happiness to receive your letter with its splendid promise of your completed manuscript for WORLD PERSPECTIVES in the visible future. I am profoundly thankful to you and await your manuscript with enduring gratitude and esteem.[11]

Your thought that Father Cantore might undertake the translation is interesting and he might indeed be the best choice. Perhaps, however, the final decision should be made, if you agree, after the manuscript is completed since as a principle it is preferable that the translator be a native of the language into which the translation is made. Nevertheless, Father Cantore is certainly well-versed in English and he might eventually prove to be the best choice although I am not able to judge his ability in the German language.

I shall be delighted to receive from you as soon as possible whatever of your manuscript is now available and I beg you to send it to me at your earliest convenience. As you know, my impatience has been boundless to have your new book in this Series and it is therefore with a great hymn of praise and appreciation that I shall await your chapters of your book as they are finished.

I beg you to let me know as soon as your plans for your American journey are settled so that we can plan a proper welcome for you.

---

11. Ruth Nanda Anshen was the general editor for the World Perspective Series at Harper & Row. The Board of Editors of the series, cited in the letter paper, was composed by: Sir Kenneth Clark, Richard Courant, Werner Heisenberg, Konrad Lorenz, Robert M. MacIver, Jacques Maritain, Joseph Needham, Robert J. Oppenheimer, Isidor Isaac Rabi, Sarvepalli Radhakrishnan, Alexander Sachs, Chen Ning Yang. A description of the purpose of the series can be found at the end of Heisenberg, *Physics and Beyond*, 249–57.

Richard and Nina Courant were with us yesterday and I told Richard of your letter and your book and he is enthusiastic.[12] He too is working on his volume, *Mathematics and the Human Mind*, for this Series. He has such a distinguished intellect and intelligence and I cherish his friendship. They both send you all warm greetings.

Again allow me to thank you for your suggestion that you send me whatever part of your important manuscript which is ready.

With enduring esteem and gratitude,
Faithfully,

Ruth Nanda Anshen

P.S. If you wish, to have your contract prepared now, I beg you to let me know. You remember that you will receive an advance royalty of $5,000.00.

---

12. In 1919, Courant married Nerina (Nina) Runge (1891–1991), a daughter of the Göttingen professor for Applied Mathematics, Carl Runge (of Runge-Kutta fame). Richard and Nerina had four children: Ernest, a particle physicist and innovator in particle accelerators; Gertrude (1922–2014), a PhD biologist and wife of the mathematician Jürgen Moser (1928–1999); Hans (1924–2019), a physicist who participated in the Manhattan Project; and Leonore (known as "Lori," 1928–2015), a professional violist.

"THE PHILOSOPHER NEEDS THE CREATIVE SCIENTIST."

Letter 44. From Enrico Cantore to Werner Heisenberg (November 28, 1968)

Heythrop College
Chipping Norton, Oxon (England)

November 28, 1968

Dear Professor Heisenberg:

Knowing your great willingness to help, I take courage in requesting your assistance. It concerns the publication of my book which—barring surprises—should take place fairly soon. The publishers (MIT Press) are interested in having as much information as possible to promote the book itself. They would like to have information about the following points, not only in the USA but also in other countries.

1) Leaders in the field (name, qualification and address) who, if given a copy of the book, would be likely to recommend it. I think that a certain number of such people are to be found in Germany. Besides Professor v. Weizsäcker, I am thinking of some of the writers of the Festschrift for your sixtieth birthday, or possibly some of your former colleagues and students. Of course, the information you give me should not put any obligation on you. The publishers would send the copy without telling the addressee who has given their name.

2) Journals which are likely (a) to review the book or (b) at least advertise it. For instance, *Universitas* or *Physikalische Blätter*? Or some others? Please, here too, name and address.

3) The name and address of any association which may be interested to bring the book to the knowledge of its members.

I thank you in advance for whatever it will be possible to you to do in this area. I hope it will not take too much of your precious time.

With most sincere thanks and the best greetings, I wish you a blessed Christmas.

Sincerely yours.

(Enrico Cantore)

Letter 45. From Werner Heisenberg to Enrico Cantore
(December 12, 1968)

December 12, 1968

Dr. Enrico Cantore
Heythrop College
Chipping Norton / Oxon
(England)

Dear Mr. Cantore.

You ask me who among the physicists or philosophers in Germany might be interested in your book and might be willing to recommend it. In addition to Weizsäcker, whom you have already mentioned, perhaps G. Süssmann (Professor of Theoretical Physics at the University of Munich), H. Dolch (Professor of Philosophy at the University of Bonn), F. Hopp (Professor of Theoretical Physics at the University of Munich), E. Scheibe (Professor of Philosophy at the University of Göttingen), and of course Father Büchel, whom I do not need to mention, could be considered.

As for journals that might consider reviewing your book, in addition to *Universitas* (Wissenschaftliche Verlagsgesellschaft, Stuttgart) and *Physikalische Blätter* (Physik-Verlag, Mosbach/Baden), you might also think of *Philosophische Rundschau* (Verlagsbuchhandlung Mohr, Tübingen), *Studium Generale*, and *Die Naturwissenschaften* (both Springer-Verlag, Heidelberg).

Regarding the third question in your letter, I do not have a definite answer. In Germany, it is customary for the most important books to be displayed by the respective publishers at scientific conferences (e.g., of the German Physical Society or the Society for Philosophy). Perhaps your publisher should consider joining this practice.

With many warm regards and best wishes for Christmas,

Yours sincerely,

H.

Letter 46. From Ruth Nanda Anshen to Werner Heisenberg (January 28, 1969)

World Perspectives—Harper & Row Publishers
14 East 81st Street
New York 28, NY

January 28, 1969

Dear Professor Heisenberg:

Fearing that my letter to you of November 21 may have gone astray, I am again taking the liberty of sending you my eager hopes that I may have the privilege of receiving your manuscript, or part of it as you so generously suggested, in the visible future.

The continuing influence of the volumes in WORLD PERSPECTIVES is most gratifying since it does indeed appear that these volumes are having an effect analogous to that of a properly oriented crystal in a saturated solution, i.e., the archetypal action of seeds is, as I understand, the same in physics as in human affairs.

Therefore your own book becomes more and more important since you yourself assure us that the spiritual and moral needs of man as a human being and the scientific and intellectual resources at his command for life may be brought into a productive, meaningful and creative harmony.

I do hope also that your plans for visiting us will soon be realized. I trust too that you will allow me to send you your contract and your advance royalties as soon as you indicate your wishes in this matter.

With every warm wish and special greetings from the Courants, believe me,
   With esteem,
   Faithfully,

   Ruth Nanda Anshen

P.S. If your manuscript should be taking the form of your Memoirs, that would be an additional value to your book since it would indeed relate and reconcile theory with your personal intellectual life and would prove to be an extraordinarily stimulating book for this Series.

Letter 47. From Werner Heisenberg to Ruth Nanda Anshen (February 3, 1969)

February 3, 1969

Mrs. Dr. Ruth Nanda Anshen
14 East 81st Street
New York 28, NY, USA

Dear Mrs. Anshen,

Please excuse my delay in responding to your letter from November. The work on the book is progressing quite well, and I am sending you a copy of the first eleven chapters, which will make up roughly half of the book. From these chapters, you can see how the entire book is structured. The following chapters will address some political misfortunes, for example, a discussion with Planck, one with Weizsäcker about the problem of the atomic bomb, a conversation with Bohr and Pauli about the issue of positivism and metaphysics, a collaboration with Pauli on the "world formula," and, finally, a discussion of atomic physics and Platonic philosophy. However, these last five chapters still need to be written, and the previous five still need editing.

Regarding the translator, I believe that Mr. Cantore's knowledge of German is sufficient for the translation, but I cannot assess the quality of his English writing style (which is certainly much better than my own). This assessment, therefore, I must leave entirely to you. Perhaps we should require that the translator not only have a command of standard English for the natural sciences but also be capable of producing a polished, artistic English style, as is customary in the humanities.

For example, I greatly enjoyed the English translation of my lecture series *Wandlungen in den Grundlagen der Naturwissenschaft* (translated as *Philosophical Problems of Nuclear Science*), which was facilitated by Eliot and published by Faber & Faber. The English style of Mr. Hayes, whom I unfortunately never met personally, seemed to me to be far superior to my own, even though occasional improvements were necessary regarding scientific terminology. I would be grateful if, after reviewing this first half of the book, you could write to me again to let me know if it meets your requirements—in which case you could also send me your contract—and if you have further suggestions for the translator.

With many warm regards
Yours

H.

P.S. I would like to give the German edition of my book, as with *Physics and Philosophy*, to a German publisher, so it should be excluded from the contract. As for translations into other languages, we would need to decide how to proceed, as it would make sense for the translations to be based on the German manuscript. Perhaps you could provide suggestions on this point.

Letter 48. From Enrico Cantore to Werner Heisenberg
(February 20, 1969)

Castello de Geronimo
Vico Equense (Napoli)

February 20, 1969

Dear Professor Heisenberg:

Sincere thanks to you for your letter of December 12, and your suggestions for advertising my book. I am glad to report that I have signed the contract for publication on December 17. The book is scheduled to appear during the next autumn. Please, accept my most sincere thanks for this event, which is so much important for me [sic]. It can help to open many doors. Certainly your generous recommendation has contributed very much to the fact that the publishers accepted so promptly my manuscript.

I had to leave England earlier than expected. A succession of circumstances compelled me to do so. As a result, I came down to Italy in order to consult my Superiors about the future of my work.

After extensive consultations, I am glad to report that the general agreement (including the personal wish of the General)[13] is that I should continue to develop the work begun. Thus I am now getting ready for going to New York again. I expect to be there by the end of March.

My immediate program is to carry out the plan of research and writing about scientific humanism I have mentioned to you in October. Then, as soon as possible, I intend also to establish contacts with academic institutions where I could develop the work while having a salary and academic standing. I am thinking particularly of the Rockefeller University.

When in New York, I shall visit Dr. Anshen. I should certainly be glad if it would be possible to me to carry out the translation of your book *Begegnungen und Gespräche*. The first reason, of course, would be the great philosophical advantage that I expect to obtain from that translation. But also the financial side of the undertaking would be welcome to me. For life in New York is very expensive, and I have no fixed income.

---

13. Father Arrupe, who asked McLaughlin to resign. Pedro Arrupe Gondra (1907–1991), SJ, was a Spanish Basque priest who served as the 28th Superior General of the Society of Jesus from 1965 to 1983. He has been called a second founder of the Society, as he led the Jesuits in the implementation of the Second Vatican Council.

My Superiors have agreed to lend me money, so that I may be free in researching and writing, but naturally I should refund it as soon as possible.

To sum up, I feel confident and desirous to develop my contribution to scientific humanism. I think it a duty of mine toward the scientific world, and also a contribution to help the Church be brought up to date in a very significant area. I have no words to thank you for the immense encouragement that you have given to me. You know how discouraged I was when I left Fordham! I thank you also because I know that I can count on your support for the future. I shall keep you informed about later developments. For the moment I do not know for sure where I shall live. But I can be reached c/o Dr. William Quiery, 106 West 56 Street, New York 10019, USA, who is the agent of my book.

Thankfully yours

(Enrico Cantore)

Letter 49. From Werner Heisenberg to Enrico Cantore (February 25, 1969)

February 25, 1969

Dr. Enrico Cantore
Castello de Geronimo
Vico Equense
Napoli (Italy)

Dear Dr. Cantore,

Thank you very much for your letter. I think it is an excellent idea for you to visit Dr. Anshen in New York. In the meantime, I have corresponded with Dr. Anshen about the possibility of you undertaking the English translation of my book, and I have confirmed that your knowledge of German is certainly sufficient for this task. The main question will therefore be whether Dr. Anshen and her collaborators consider your command of the English language to be equally suitable. This is something I cannot assess, as my own English is not good enough for such an evaluation. We will need to wait and see how this develops. I would be grateful if you could keep me informed about your discussions with Dr. Anshen.

With best wishes,
Yours sincerely,

H.

Letter 50. From Ruth Nanda Anshen to Werner Heisenberg (March 7, 1969)

World Perspectives—Harper & Row Publishers
14 East 81st Street
New York 28, NY

March 7, 1968

Dear Professor Heisenberg:

What a splendid manuscript you have written, even more compelling, if such a thing could be possible, than your distinguished *Physics and Philosophy*.

 I have finished reading it and it is now at Harpers having the same enthusiastic reaction. For you have pointed to the meaning of Order in the very nature of science, and your autobiographical structure of your fine book lends intimacy and experience to your text. It seems to me that the postulate of "order" may be regarded as one of the highest-level hypotheses in every field of science. For you show that neither in physics nor in any other field of science can scientific generalizations be established for an erratic universe. And the significant questions revealed from your text compel us to ask: how can the open structures prevailing in the life sciences be related to the rigid order prevailing in physics and chemistry? Or can positive insights into the world of Man be obtained independent of normative judgments? Or again, what are the implications for the ontological principles underlying the scientific approach for the relationship between knowledge and action, for the unity of science?

 Your volume in WORLD PERSPECTIVES, your companion new volume to *Physics and Philosophy*, your *Encounters and Dialogues*, is simply magnificent. I am so pleased and so proud, and thank you profoundly. As I understand, there will be more chapters to follow. May I beg you to let me know when to expect them? That is, if you feel that you want to add more chapters?

 Harpers is preparing your contract which I shall be sending to you with great honor and gratitude. Upon your signature your royalties will be sent to you.

 With the warmest wishes.

Faithfully,

Ruth Nanda Anshen

P.S. I have noted your suggestion about Mr. Hayes as translator and have written for a copy of your *Philosophic Problems of Nuclear Physics*.

Letter 51. From Enrico Cantore to Werner Heisenberg
(March 22, 1969)

New York, March 22, 1969

Dear Professor Heisenberg:

I thank you for your letter of February 25. As soon as possible I shall visit Dr. Anshen and discuss with her my possible contribution to the translation of your book. I shall write to you about her decision at the earliest possible date.

With this letter I wish to inform you about the present situation at the Rockefeller University and the possibilities that can be available for me there. A friend of mine, who is a professor at RU and has carefully followed the development of events there, has given me the following picture. RU finds itself in a state of transition and a certain difficulty because of two main reasons. The first is a state of unrest and criticism against the administration on the part of the junior professors and students due to the mentality currently prevalent among the younger generations. The second reason is the financial strain that the University must undergo because of its building expenses and the growing demands of its technicians and junior professors. Despite these difficulties, however, my friend has suggested that it is reasonable for me to submit to the President, Dr. Seitz,[14] my request for working there toward the development of my program. In particular, my friend suggested that the present is the best suited time of the year for making such an application. For it is during March and April that the budget of the next school year is being prepared. Consequently, I modestly beg you to be so kind as to present me to the same President.

I summarize here the reasons why I wish to be associated to the academic community of RU. The central motive is my desire to contribute effectively to the solution of the so called two-culture problem. It seems to me that RU has a special interest in this field. The purpose of the University, as officially stated, is the furtherance of science and its application for the improvement of human welfare. In actual fact,

---

14. Frederick Seitz (1911–2008) was an American physicist, succeeded Bronk as fourth president of Rockefeller University from 1968 to 1978. He was also the seventeenth president of the United States National Academy of Sciences from 1962 to 1969. Seitz was the recipient of the National Medal of Science, NASA's Distinguished Public Service Award, and other honors.

RU has increasingly tried to foster the synthesis of the sciences and the humanities. This is shown by the fact that the University has gradually added the natural sciences, mathematics and philosophy to the original field of research which consisted almost exclusively of medicine and biological sciences. As a consequence, it is likely that the University may be interested in my specific contribution. For my aim, as you know, is precisely that of bringing about a more adequate conception of man, and of his values and ideals, in the light of the deeper perspectives and insights offered by modern science.

The complementary reason why I wish to join the Rockefeller University is the fact that this University would offer me a unique opportunity for developing my program. To achieve my aim, I need a continual and extensive dialogue with creative scientists of various fields. But such a dialogue, as I know from experience, is practically impossible in ordinary universities. For there the scientists are rigidly organized in specialized departments. RU, on the contrary, is built around individuals rather than departments. This unique organization would make the inters disciplinary dialogue I have in mind very fruitful for me.

As for the practical form of my association with RU, I am perfectly willing to leave the matter in the hands of the President. My central desire is that of being given the possibility of personal and frequent encounter with the scientists of the University and of being allowed the use of its educational facilities (such as libraries, lecture series, etc.). In view of the special difficulties that the President has to face at present, I do not ask to be paid a regular salary. If possible, I would certainly be pleased if I could be given some little money to cover my living expenses. But I have no intention of insisting even on that.

To conclude, I am very thankful to you for your great understanding and generous encouragement. For my part, I intend to do everything in my power in order to serve man in the scientific age. I shall not stop trying until circumstances may show that the attainment of my aim is impossible.

As soon as I shall know that you have introduced me to Dr. Seitz, I shall write to him myself, specifying the program I have in mind, (in case you think that a copy of your letter of presentation to the President may be useful to me, best thanks also for that. But, please, feel perfectly free in this regard.)

May the Lord reward you abundantly for all the good you are doing me!

Thankfully yours,

(Enrico Cantore)

Here is the complete address of Dr. Frederick Seitz, President, Rockefeller University, New York 10021 USA

—Since I am not yet sure where my permanent address will be, please address your correspondence to me still c/o Dr. William H. Quiery, Jesuit Writers' Service, 106 West, 56 Street, New York 10019, USA

Letter 52. From Werner Heisenberg to Ruth Nanda Anshen (March 25, 1969)

March 25, 1969

Mrs. Dr. Ruth Nanda Anshen
14 East 81st Street
New York 28, NY, USA

Dear Mrs. Anshen,

Thank you very much for your letter. I was pleased to see that you could work with my book, and your feedback encourages me as I work on the final chapters. I still have three chapters left to write, and I believe I will be able to finish the entire project by May.

Regarding the translator, I will, of course, leave the decision largely in your hands. I only mentioned Mr. Hayes's translation because it seemed stylistically excellent to me; however, I do not know whether Mr. Hayes is still alive or still undertakes such work. I would be happy if the translation could be entrusted to Father Cantore, as he deserves support and encouragement, even aside from our specific project. His knowledge of German is certainly sufficient, but I must leave the evaluation of his English skills entirely to you, as you have a much better and more reliable judgment in this regard.

I have not yet received a draft contract from Harper & Row. I will likely only be able to respond to a letter from Harper after returning from my Easter trip. I will be leaving here in a few days and will not return to Munich for an extended period until early May.

With warm regards, also to Mr. and Mrs. Courant,

Yours sincerely,

H.

Letter 53. From Werner Heisenberg to Frederick Seitz, Rockefeller University (March 28, 1969)

March 28, 1969

Dr. Frederick Seitz
President, Rockefeller University
New York, NY 10021, USA

Dear Seitz:

Dr. Enrico Cantore has given me a note that he is intending to apply for a post at the Rockefeller University, and he has asked me to send you a few lines of recommendation.

I have known Dr. Cantore for a few years since we had some discussions in my institute on his book *Atomic Order—An Introduction to the Philosophy of Microphysics* and on general aspects of the relation between old humanistic philosophy and modern science. By our conversations and by several letters we exchanged, I got the strong impression that Dr. Cantore takes a profound interest in the problem of a synthesis of the sciences and the humanities; by his thorough education and his very wide knowledge he is certainly well prepared for doing work in this field. To show my confidence in his abilities I should like to mention that I suggested Dr. Cantore for translating my latest book dealing with physical and philosophical aspects.

I would consider it a great advantage for Dr. Cantore to join your university in order to obtain a possibility for pursuing research and to have frequent opportunity for discussions with the scientists of your university. Therefore I should like to support Dr. Cantore's application in every way.

With best regards
Yours,

H.

Copy to Dr. Cantore

Letter 54. From Ruth Nanda Anshen to Werner Heisenberg
(March 29, 1969)

March 29, 1969

World Perspectives–Harper & Row Publishers
14 East 81st Street
New York 28, NY

Dear Professor Heisenberg:

I was about to send your contract to you which I now enclose with so much honor and gratitude when your letter of the 25th arrived.

Yes, your manuscript is like an intellectual, scientific, moral and spiritual thermostat, giving the reader a magnificent canvas, like a symbolic painting of the twentieth century. It is indeed a special privilege to include it in WORLD PERSPECTIVES.

I shall await the pleasure of receiving the last three chapters for your volume sometime during the coming May, as you generously suggest.

Father Cantore is in New York now and will come to see me next Tuesday. I shall discuss the possibility of the translation of your book by him, at least tentatively since I am a little concerned about the adequacy of his knowledge of English. I shall of course report to you the results of my conversation with Father Cantore.

Richard Courant is on his way to Japan to receive many honors . . . at the age of 82! He is in wonderful health and vigor, and he said that your manuscript is most important and so necessary for our time. Professor and Mrs. Courant send you their warm greetings.

With most enduring appreciation and cordial wishes,
Faithfully,

Ruth Nanda Anshen

Enclosures

P.S. You will find herewith enclosed two copies of your contract already signed by Harper & Row. Please keep one copy of the contract for your own files and return the other copy to me, after the receipt of which I shall instruct the publisher to send you a check for $5,000.00.

Letter 55. From Enrico Cantore to Werner Heisenberg
(April 2, 1969)

New York, 2 April 1969

Dear Professor Heisenberg:

Only yesterday was Dr. Anshen able to grant me an interview. I hasten to report to you about the present state of affairs concerning the translation of your book into English.

Dr. Anshen was not in a position to take the final decision. She mentioned to me the objections of principle that militate against choosing as a translator a person for whom English is not the native tongue. On the other hand, she preferred to examine the issue further. Shen [sic] asked me to give her a text written by me in English in order to evaluate my linguistic ability. In addition, she plans to discuss this problem with some expert of the publishing firm Harper & Row before taking the final decision.

As for me, I understand perfectly well this situation. Therefore, while acknowledging my limitations, I intend to keep myself available. But I do not wish to interfere in any way with a decision which is important and involves so many delicate aspects.

I think that Dr. Anshen herself will write to you as soon as possible once the issue has been settled. For my part, I promise to keep you informed in case I would be invited to undertake the translation. I wish to you and your work, especially to this wonderful book, the best of success.

Probably you have already received my letter of March 22. Please have my best thanks for this and all the immense and most generous help you have given me.

Sincere paschal greetings to you and your family.

Sincerely yours.

(Enrico Cantore)

Letter 56 from Frederick Seitz, Rockefeller University, to Werner Heisenberg (April 9, 1969)

The Rockefeller University
New York, NY 10021
Office of the President

April 9, 1969

Professor Werner Heisenberg
Max Planck Institut Für Physik Und Astrophysik
München 23, Germany

Dear Professor Heisenberg:

Thank you very much for your letter of March 28th regarding Dr. Enrico Cantore. I would be pleased to review his application in the coming weeks. In examining our records I find that he was considered by the faculty about a year ago and the conclusion was drawn that the opportunity was then not ripe. Let me see how matters stand at the present time.

Sincere regards,

Frederick Seitz
President

Letter 57. From Enrico Cantore to Werner Heisenberg
(April 16, 1969)

April 16, 1969

Dear Professor Heisenberg:

With these few lines I intend to acknowledge the reception of a copy of your letter of recommendation in my behalf to the President of the Rockefeller University. Immediately upon receiving your communication I wrote to Professor Seitz applying for a position at his University.

It is both a duty and a pleasure for me to take advantage of this occasion to express my most sincere and heartfelt gratitude to you. May the Lord reward your great gentleness and generosity!

For the moment I have no news of significance to communicate, I shall dutifully keep you informed about every future development.

With best greetings,

(Enrico Cantore)

Letter 58. From Joan Nelson Warnow, American Institute of Physics, to Werner Heisenberg (April 18, 1969)

American Institute of Physics
335 East 45 Street, New York, New York 10017   Murray Hill 5-1940
Center for History and Philosophy of Physics
Charles Weiner, Director

April 18, 1969

Professor Werner Heisenberg, Director
Max Planck Institut für Physik
8 München 23
Föhringer Ring 6
Germany

Dear Professor Heisenberg:

Father Enrico Cantore has applied for access to the manuscript materials in the Niels Bohr Library. He is investigating the humanistic meaning of science, i.e., the new conceptions, perspectives and problems brought about by science concerning man, his relationships to reality, his ideals and values. According to our procedures. Father Cantore has been asked to give the names of two references. He has listed you as a reference.

It is the policy of the Center for History and Philosophy of Physics to make manuscript materials available to qualified researchers who will use them responsibly. I would appreciate a brief comment from you indicating whether or not you feel Father Cantore meets this standard.

Sincerely,

(Mrs.) Joan Nelson Warnow
Librarian

Letter 59. From Werner Heisenberg to Joan Nelson Warnow, American Institute of Physics (April 24, 1969)

April 24, 1969

Prof. W. Heisenberg

American Institute of Physics Library
Mrs. Joan N. Warnow
335 East 45th Street
New York, NY 10017, USA

Dear Mrs. Warnow:

In reply to your letter of April 18th I should like to support in every way Father Cantore's application for access to the manuscript materials in the Niels Bohr Library. I know Father Cantore as a qualified researcher and I am convinced that he will only use the manuscript materials in that responsible way you would expect of him.

    Sincerely yours.

    H.

Letter 60. Telegram from Ruth Nanda Anshen to Werner Heisenberg (May 8, 1969)

May 8, 1968

BEG YOU RETURN YOUR CONTRACT WORLD PERSPECTIVES WITH YOUR SIGNATURE SENT YOU MARCH 29 GRATITUDE GREETING ANSHEN

Letter 61. From Werner Heisenberg to Ruth Nanda Anshen (May 9, 1969)

May 9, 1969

Mrs. Dr. Ruth Nanda Anshen
14 East 81st Street
New York 28, NY

Dear Ms. Anshen,

Thank you very much for your letter and the draft contract with Harper & Row. In this draft, two wishes I mentioned, if I recall correctly, in one of my previous letters, have not been addressed. I would like, similar to the arrangement for *Physics and Philosophy*, to retain the rights for the German edition and transfer them to the Piper Verlag here in Germany. I assume Harper & Row will not have any fundamental objections to this, as the same approach was taken with the earlier book *Physics and Philosophy*. This leaves the question of whether the publisher insists that the German edition not be released before the English edition. It is very important to me that the German edition appears by the end of this year at the latest, and I am willing to make every effort to ensure that the English edition is completed by then as well. However, this largely depends on the English translator, and I do not know whether you have already found a suitable translator or how quickly the work will proceed. If translation delays occur, I would think that the earlier release of the German edition would not significantly impact the sales of the English edition, especially since promotional materials for the English edition could be distributed earlier. I would like to hear your opinion on this point.

Additionally, I noticed that the publisher wishes to reduce the royalty for the first 10,000 copies to 7.5 percent of the retail price (compared to the 10 percent agreed upon for *Physics and Philosophy*). I assume this is due to the cost of translating the book into English. As compensation, the royalty for subsequent copies has been set at 12.5 percent. I would appreciate it if these terms could be improved slightly, for example, by increasing the royalty for later copies to 15 percent or, if the publisher prefers, by raising the royalty for the first 10,000 copies.

Finally, I find the proposed title of the German book (*Begegnungen und Gespräche* [Encounters and Conversations]) a bit bland. I would prefer to rename it *Gespräche im Umkreis der Atomphysik* (*Conversations in the Realm of Atomic Physics*). The title for the English edition should be decided in consultation with the translator. The length of the German manuscript will be about 285 pages, and this figure should also be correctly stated in the contract.

The book is now nearly complete; only the final third of the last chapter remains. I hope to send you the complete manuscript for translation by the end of next week.

Regarding the final version of the contract, I could either incorporate the desired changes myself and sign it immediately or wait for Harper & Row to send me the finalized version of the contract. Please let me know how I should proceed.

I also thank you and the publisher for the generous offer to send me an advance of $5,000. Once the legal arrangements are in place, I would appreciate receiving the check promptly, as it would naturally be advantageous for me to receive it before a possible revaluation of the Deutsche Mark.

Once again, thank you very much for your letter and your efforts. I apologize for only replying now, after returning from an extended trip to the Mediterranean.

With many warm regards,
Yours sincerely,

H.

Letter 62. From Ruth Nanda Anshen to Werner Heisenberg (May 13, 1969)

World Perspectives–Harper & Row Publishers
14 East 81st Street
New York 28, NY

May 13, 1969

Dear Professor Heisenberg:

I have today received your detailed letter concerning your important book which honors WORLD PERSPECTIVES.

The first paragraph relating to the German rights which you prefer to be handled by you and given to Piper-Verlag, this is completely acceptable and I beg you to make this change in your contract which you now have and place your initials beside this change in the margin of your contract, i.e., in both copies keeping one for your files and returning the other copy to me.

If you should agree, it would be desirable for Harper & Row to continue to give the paperback rights of the German edition to Ullstein Verlag who publishes the volumes in WORLD PERSPECTIVES. If, however, you should consider this to be impossible, then of course Harper & Row will agree to give you the full rights, hard cover and paperback, for the German edition of your book. Please place your initials beside this change in the margin of your contract, in both copies.

As soon as you send me the remaining chapters of your book, the entire manuscript will be given to A. J. Pomerans, an excellent translator. When he has the complete manuscript he will be able to judge the time required before finishing the translation. Your book should be published in all probability early in 1970, or at the end of 1969, if the translation is completed in time.

Harper & Row agree to your terms that you receive 7.5 percent up to the first 10,000 copies and 15 percent thereafter. Please place your initials after you make this change in your contract.

Concerning the title of your book, I agree that it should be changed and perhaps it would be best to wait until the English translation is completed before making the final decision. I shall take the liberty of sending you my suggestions and shall await your own decision in this matter.

Concerning the number of pages, I beg you to make whatever change you feel is reasonable in your contract and simply state that there will be *approximately* the number of pages you wish to have. The page numbers will of course be different in English from what they are in German.

I shall await the remaining chapters at the end of the week, as you so generously promised, and as soon as you return your signed contract with your signature, I shall instruct Harper & Row to send you $5,000.00. It appears that the German mark will not be devaluated after all.

I hope, with all my heart, that all your requirements have now been fulfilled and beg to thank you again and again for the profound privilege of including your important and distinguished volume in this effort. Through your extraordinary intellectual powers and spiritual and moral vision, the sorcerer's apprentice may still be outwitted.

With all my warmest thoughts,
Faithfully,

Ruth Nanda Anshen

P.S. Please attach the enclosed clauses to each copy respectively of your contract and initial them RNA.

Letter 63. From Werner Heisenberg to Ruth Nanda Anshen (May 14, 1969)

May 14, 1969

Mrs. Dr. Ruth Nanda Anshen
14 East 81st Street
New York 28, NY, USA

Dear Mrs. Anshen,

I am sending you the last nine chapters of my book by the same post. The text is now complete, but before the manuscript is handed over to the translator, I would like to make a few small revisions, which will only involve minor stylistic improvements and similar adjustments.

As soon as you inform me that a translator has been found, I will send you or the translator a finalized copy.

With many warm regards,
Yours sincerely,

H.

Letter 64. From Werner Heisenberg to Ruth Nanda Anshen (May 19, 1969)

May 19, 1969

Registered Mail
Mrs. Dr. Ruth Nanda Anshen
14 East 81st Street
New York 28, NY, USA

Dear Mrs. Anshen,

Thank you very much for your letter. I have made the changes to the contract as you suggested and am enclosing the signed contract with this letter. Regarding the title, I have noted in the margin that changes can be made by mutual agreement. I hope this resolves the legal prerequisites for the English edition.

I was sorry to hear that Father Cantore did not quite pass your evaluation, but of course, it is up to you to assess his command of the English language. I am not yet familiar with Mr. Pomerans; has he already translated scientific or literary works, and does he have some familiarity with philosophical terminology? I assume you have already discussed these questions with him. Hopefully, he will not take too long to complete the translation, so the book can also appear in English by the end of the year.

You also inquired about paperback rights for the German edition, which you would like to assign to Ullstein Verlag. I believe it would not be reasonable to propose such a clause to Piper Verlag, and I kindly ask for your understanding that I intend to transfer all rights for German editions to Piper Verlag.

Since I want to make a few small improvements to the text over the next few days, I will send you the finalized version of the manuscript within the next ten days. However, the translator can begin his work now, as these will only involve minor stylistic adjustments.

With this, I hope all legal questions are settled, and work on the publication can begin.

With many warm regards, also to Mr. Courant,
Yours sincerely,

H.

Attachment: Contract

"THE PHILOSOPHER NEEDS THE CREATIVE SCIENTIST."

Letter 65. From Enrico Cantore to Werner Heisenberg
(May 31, 1969)

May 31, 1969

Dear Professor Heisenberg:

It is my duty to inform you about the answer recently received from Dr. Seitz, the President of Rockefeller University. After a very long delay, and after having had time to examine some writings of mine, he said that he had "reluctantly" come to the decision that he could not give me "an appropriate position" there.

Needless to say, this has been a severe blow to me. All I had asked for, was an unpaid researcher position which would have enabled me to be in a scientific environment, and given me the possibility of a dialogue with working scientists.

I do not know the reasons for this second refusal. One thing seems certain, however; the philosophers at Rockefeller are those who raised the real opposition against my being accepted. I know this for sure, through a professor of the University: nor am I surprised, given the mentality of the philosophers there who seem to belong wholly to the analytic-logistic tradition.

It is discouraging to see how the present institutions, despite the occasional high-flying rhetoric about the intercultural dialogue, are so unbending and un-cooperating when someone tries to bring this dialogue about. On the other hand, I think it is my duty not to refuse my personal contribution toward making this dialogue a reality. Hence I have decided to do whatever possible in this direction. I plan to remain in New York concentrating particularly on writing and also organizing private meetings among professors of various universities interested in the dialogue itself. If we will be able to find the right formula, I think we will then put it on a permanent basis and make it public through series of open lectures, publications, etc. I shall keep you informed about all developments. The more I realize how little concern people show for the intercultural dialogue, the greater becomes my gratitude and respect for you who have really done everything possible to help me. In this connection, please allow me to express my confidence in you also concerning the future. I shall try my best not to inconvenience you in any way. But, in case that circumstances would indicate that your coming to New York may help make a success of the dialogue I am now trying to develop, I think I should bring this request to you.

I am now doing research for a book on the humanistic significance of science,[15] the first of a series of five. I know from Dr. Anshen that your manuscript of *Begegnungen und Gespräche* is finished.[16] It would certainly be of immense help to me in planning my book if I could have soon a copy of your book (or some preprint form of it, if possible). Please let me have some information about when I could have the text of your book. As for doing the translation. Dr. Anshen must have given you all the information available. The publishers themselves. Harper & Row, are apparently taking care of the matter.

With most heartfelt thanks, and respectful greetings,
I am sincerely yours,

(Enrico Cantore)

---

15. This is the first mention of Cantore working on a manuscript that will later become the book *Scientific Man* (1977). Heisenberg will support this project, without seeing its publication: the Nobel laureate will pass away in early 1976.

16. This is the German subtitle of *Der Teil und das Ganze*.

Letter 66. From Werner Heisenberg to Enrico Cantore
(June 6, 1969)

June 6, 1969

Dr. Enrico Cantore
c/o Dr. William Quiery
106 West 56 Street
New York, NY 10019, USA

Dear Dr. Cantore:

I am sorry to hear from your letter about the great difficulties you have encountered at Rockefeller University. I suspect, as you do, that these are related to the conflicts between various philosophical directions, and for now, it seems the logicalist direction is dominant.

I was also very saddened to learn that Dr. Anshen and Harper & Row decided on a different translator for my book. Naturally, I had to leave this decision to the Americans, as I am not confident enough in the English language to make a judgment here. I have not yet received detailed information or sample texts from the translator they have now chosen.

I hope you will not lose heart in what you have set out to do, and if I can help you in any way, I will gladly do so; but you can see that my possibilities are very limited.

My book will be published in German by Piper Verlag toward the end of the year. How long the English translation will take remains to be seen. I am sending you a copy of the typed manuscript by the same post but kindly ask you to treat the manuscript confidentially, i.e., not to share it with others or quote from the text before it is published.

With best regards,
Yours sincerely,

H.

IV.

"You tell me not to lose courage."

Continued Support and Publication of
"Physics and Beyond," June 1969–May 1970

CANTORE'S WORK IN NEW York, characterized by precarious finances and institutional troubles, could not be sustained by simple intellectual excellence; it needed pragmatic backing. As this correspondence between the two men illustrates, Cantore's bid for even modest support from the American Philosophical Society owed much to Heisenberg's letter of recommendation.

In a letter of August 11, 1969, Cantore tactfully requested from Heisenberg his support for issuing a $1,500 grant by the American Philosophical Society. It was practically motivated by his research: he needed some basic office equipment, some essential books, and money to travel and interview scientists. Indeed, Heisenberg's prompt and positive response to the Society, dated August 20, listed Cantore's qualifications and the importance of his work. According to Heisenberg, Cantore is seriously interested in linking and combining sciences with human studies. Considering his deep education and vast knowledge, he figured he was well qualified for the projected work. Furthermore, he also showed his confidence in Cantore by mentioning that he suggested him as a translator for his new book.

Heisenberg endorsed Cantore, and finally, in October of 1969, Cantore got a confirmation of the $1,000 grant from the Society.[1] It was a smaller sum than he had initially applied for, but he felt it was very much in his favor. Writing on October 28, 1969, Cantore showed deep gratitude to Heisenberg and underlined that the recognition through the American Philosophical Society was a financial boon and a moral spur concerning the value of his work in a country where he was so far quite unnoticed.

Amidst these practical details, another, more important detail of their collaboration was the publication of Heisenberg's book, *Physics and Beyond*, described as an autobiographical and philosophical work (in German, *Der Teil und das Ganze*). This text exemplifies the philosophical context of quantum theory and Heisenberg's reflection. It was an important intellectual exercise related to Cantore's intellectual work. When he first encountered Cantore, Heisenberg regarded him as a potential translator for the English edition of his book. This would have been an organic offshoot of Cantore's work, since he really handled well the philosophical issues involved in quantum physics.

Despite Heisenberg's suggestion, Harper & Row, the American publisher, did not choose Cantore. If one reads a detailed comparison of letters swapped between Heisenberg and Cantore going as far back as January 1969, Heisenberg said that he regretted it very much and alluded to the fact that he had some confidence in Cantore's knowledge of German. However, the publisher appears concerned about the tone the book should take in English. At any rate, this should not have made Heisenberg underestimate Cantore's mind. When *Der Teil und das Ganze* was published in the fall of 1969, Heisenberg sent him a copy with a dedicatory note, warm with affection, that reminded Cantore of the great intellectual adventure they had undertaken together.

In his letter of January 17, 1970, Cantore recognizes how much support Heisenberg had given him and how helpful their conversation was for his work, notably his preparation for a book on the humanistic relevance of science. With this type of profound respect, Cantore had the idea, and he intended to dedicate this upcoming book to Heisenberg. It goes to say Heisenberg was not an ordinary physicist in the eyes of Cantore; he was an ideal mentor, and as a friend, he was perfect; in other words, he came up to be a true "scientist-humanist."

---

1. The endnote to Cantore's "Science as Dialogical Humanizing Process" (1971) gratefully acknowledges the contribution of American Philosophical Society for the expenses for the article.

Atomic Order also came out in 1969 by MIT Press. The profound influence of Heisenberg on Cantore's work is eloquently echoed in the preface of *Atomic Order*:

> Above all, I am indebted to Professor Werner Heisenberg. He not only read the manuscript and gave valuable suggestions, but he generously assisted this writer in many practical ways, concerning both the publication of this work and the promotion of the scientific-philosophical dialogue. Without his encouragement and assistance, it would not have been possible to me to overcome the many difficulties which threatened repeatedly to destroy my courage.

As Cantore states in a series of letters, without Heisenberg's encouragement and support, the obstacles that stood in the way of his diminishing courage may have won out. Of note is that, in another letter, Cantore mentions how Heisenberg influenced his work like Bohr had influenced Heisenberg's own.

Despite defeats, such as those involved in his attempt to translate *Physics and Beyond*, Cantore was indomitable in his drive to develop and propagate scientific humanism. He continued to canvass Heisenberg for advice and assistance on many fronts, including planning an English translation of an Encyclopedia of Philosophy to break the hegemony of positivist philosophy in English-speaking nations. However, after Heisenberg politely declined to take up any new projects due to his already overwhelming commitments, Cantore realized that he would have to start looking for the extra help he required outside Munich.

By the middle of 1970, Cantore had been well on his way to working on the preparatory stages of what should have been his second book, to have completed the final manuscript within a year. This book is dedicated to Heisenberg and thus gives witness to the physicist's permanent impression on Cantore's intellectual development and his work on scientific humanism.

Letter 67. From Enrico Cantore to Werner Heisenberg (June 27, 1969)

June 27, 1969

Dear Professor Heisenberg:

Please excuse me for having delayed answering your kind letter of June 6 until the present. I was quite busy during this time. I had to change residence, and correct the galleys of my book which had to be returned to the publisher quite soon.

I am extremely grateful to you for your encouraging words and the generous offer of help, if circumstances would offer any opportunity for it. I was really moved in receiving so promptly copy of your manuscript book *Gespräche im Umkreis der Atomphysik*.[2] You know how much I appreciate this work of yours. To me it is the most convincing evidence of the profoundly humanistic significance of science.

You have certainly a right to be listened to when you tell me not to lose courage. Frankly, I do not know whether I would ever have found confidence to go on in the present work toward a scientific-humanistic synthesis had it not been for your continual understanding and open-hearted support. Of course, I am not able to thank you in any adequate way. As an expression of my esteem for you, I shall engage myself thoroughly in this work.

My present plans are as follows. To help pay my daily expenses I am now living in a parish where I work part-time as an assistant pastor (Kaplan). This hopefully will leave me abundant free time which I intend to dedicate chiefly to writing the series of books whose general outline I discussed with you during my October stay at your Institute.[3] Another occupation will be some teaching. The group called The Scientists for Public Information has invited me to lecture in a sort of popular university at the New School for Social Research during the coming first semester of 1969/1970. I will give a course discussing the philosophical significance of science. Finally I shall try my best to organize a dialogue of scientific-humanistic type among interested University professors of the New York area.

---

2. Later published as *Der Teil und das Ganze*.
3. Cantore is referring to his previous visit to Heisenberg in Munich.

I shall study and enjoy thoroughly the magnificent book whose manuscript you have sent me. At the present I think I am in a better position to understand it for I have followed your suggestions and have read your interviews for the History of Quantum Physics kept in the Library of the American Institute of Physics. I have taken some notes from those interviews. In addition, there are longer passages which I would like to have. The Librarian at the American Institute of Physics said she was not empowered to give me photocopies of those passages. As a consequence, I write to you for obtaining the necessary permission to quote from your interviews and also for being allowed to have photocopies. The longer texts I am interested in are as follows. From the interview marked 11/30/62 I would like to reproduce p. 14. From interview marked 11/II/63 I would like to have pp. 16/17/18. From 13/II/63 p. 7. From 2/25/63: pp. 4f. and pp. 16f. From 2/27/63: pp. 19/20/21 and p. 26. From 28/II/63: p. 22 and pp. 30f. From 5 July 1963: p. 10. If it could be possible, I would certainly be glad if you have your secretary make a photocopy for me of these pages from the final manuscript (a copy of which I read at the American Institute of Physics). But I do not want to take advantage of your goodness. Please, at least let me have your written permission to quote from those pages, plus some short quotations here and there.

By reading your interviews I discovered what I consider a most important writing of yours, namely your article on Pauli's philosophy. Unfortunately this article is contained in a journal which is not easily accessible. I hope that the English translation is a faithful one, at least. Please, tell me your opinion. Here, too, if it were possible to have a reprint, I would be most grateful, (if the English translation would not be good, I would prefer the German original.)[4]

Finally, I am interested in your work entitled: "Natural Law and the Structure of Matter," from *Frontiers of Modern Scientific Philosophy and Humanism*, Athens Meeting 1964 (Elsevier 1966).[5] This book, too, I was unable to find until now.

I have taken the liberty of mailing to you, under separate cover, two articles of mine.[6]

---

4. See Heisenberg, "Wolfgang Paulis Philosophische Auffassungen."

5. This paper was later published separately as Heisenberg, *Natural Law and the Structure of Matter*.

6. Cantore is likely referring to the same papers cited in Letter 68, Cantore, "Scientific Humanism and the University"; "Science and Humanism."

Before closing, still a little piece of news. The Director of the MIT Press, having heard of my plans of writing a series of books on the relationships between science and philosophy, asked me recently for a statement covering my plans. I wrote to him the enclosed letter. I enclose this copy here simply because in it I mentioned your name, so I think you would prefer to be informed about its contents. With the renewed expression of my most sincere gratitude and respect,

I am truly yours,

(Enrico Cantore)

Enrico Cantore
627 East 187 Street
Bronx, NY 10458 USA

Attachment: letter to the director of MIT Press, Dr. Bowen

June 11, 1969

Mr. Carroll G. Bowen, Director The MIT Press
MIT
Cambridge, MA 02142

Dear Mr. Bowen:

I thank you for the kind interest you have taken in my work. With this letter I intend to comply with your request of May 27. I shall outline the general approach and plan of my proposed contribution to the intercultural dialogue.

My starting point is the realization of the reasons causing the split between the two cultures. These reasons are fundamentally two: a mutual misapprehension and the lack of a common basis for dialogue. Accordingly, I think that my contribution, as a philosopher, should consist in helping to remove the misunderstanding and indicate the way toward a mutual cooperation. As a consequence, my projected work can be divided into two main parts. To give them a name, I would call the first part "Introduction to Scientific Humanism," while the second could be named "Principles of Scientific Humanism." Scientific Humanism is used here to express an up-to-date conception of man and reality in which the humanistic contribution of science is harmoniously integrated.

The intercultural misapprehensions can be briefly characterized as follows. The philosophers tend to misjudge science by considering it as a purely logical scheme or as a collection of practical recipes. The scientists tend to view philosophy as irrelevant or at least suspect by concentrating unduly on the weaknesses of philosophers. Consequently, I think that the most fruitful introduction to scientific humanism may be that of presenting the vital significance of both science and philosophy for a satisfactory humanism. Accordingly, I plan to write two books by way of general introduction. They can be tentatively entitled: *The Humanistic Significance of Science* and *The Philosophical Problem of Man in the Age of Science*.

The obvious difficulty at this point is that of finding a suitable method of investigation. This presents a difficulty because my aim is to honestly portray the different mentalities of science and philosophy while at the same time appealing to the interests of both scientists and philosophers. To meet this difficulty I think the most promising method consists in developing systematically the approach I have come upon while reflecting on the philosophical significance of atomic physics. I call this approach inductive-genetic. The method is inductive because it starts out from concrete data rather than general assumptions; it is genetic, because it tries to understand its subject matter by studying its development in time. In other words, while studying either science or philosophy, my first concern is that of capturing their typical spirit as different humanistic experiences involving the whole man. Then, and only then, I intend to turn to the reflective stage. This consists in analyzing the new perspectives and problems that either discipline raises for man, and the complementarity of both science and philosophy for the development of a new humanism.

In the light of the preceding, I can give here a brief outline of *The Humanistic Significance of Science*. In the first part of this book, I plan to recapture the spirit of natural science as an experience of reality which involves the whole man, both individual and social. This I intend to do by examining concrete science, especially physics. The reason for a special emphasis of physics is partly personal (this is the science I was trained in), partly historical and psychological (physics was the first experimental science to be developed in history and still remains in some sense the ideal of science as such). To capture the humanistic experience of science I shall rely mainly on biographical and autobiographical accounts of leading scientists. In addition, I shall systematically investigate the histories of the different branches of experimental science. Finally, I shall study the principal philosophical writings dealing with my subject matter, especially those composed by creative scientists discussing their scientific experience.

In the second part of the book, I intend to examine the new philosophical perspectives and problems arising out of science viewed as a lived experience of the total man. My intention in this book is not that of solving such problems, but simply of stating them clearly. I hope that honest-minded philosophers will realize what a tremendous challenge is offered them by modern science, and how greatly philosophy itself can progress if it can be persuaded to take science seriously into account.

As for *The Philosophical Problem of Man in the Age of Science*, here too I plan to divide my book into two main parts. In the first part I plan to examine the historical relevance of philosophy in trying to solve the problem of man. This I intend to do by following the main stages of modern philosophy, starting from the humanistic crisis that broke into the Western world at the rise of modern science, and following the development of philosophy up to our times. I shall focus my attention on the most significant authors and currents, with the intent of pointing out their main positions, and the reasons thereof.

In the second part, I shall turn to a reflective analysis of the results of philosophy in the age of science. I intend to emphasize the achievements and progress of philosophy. At the same time, however, I shall endeavor to discover the reasons that are responsible for the unsatisfactory state of modern philosophy when confronted when the problems of scientific man. In short, here too, my intention is not that of solving problems, but merely that of analyzing and stating issues in a clear and convincing way. Thus I hope that honest-minded scientists will realize the relevance of philosophy and will join their efforts to those of the philosophers in order to help man find self-understanding in the age of science.

I can now give you an idea about the second part of my programmed work. I have called it "Principles of Scientific Humanism" because in it I intend to present some fundamental principles which should serve as guidelines for the intercultural dialogue aimed at developing a satisfactory scientific humanism. The subdivision of this part of my projected work follows immediately from the nature of philosophy. The main question of philosophy, as well known, can be reduced to problems of epistemology, ontology, and ethics. Accordingly, I intend to write three different books, one for each of these subjects. As for the general methodology that I intend to adopt, I would characterize it as a conscious effort of formulating a philosophical view of man and reality which takes science seriously into consideration. Of course, this work of mine is intended to be no more than a tentative synthesis, always open to suggestions and criticisms. In other words, I aim at providing an example of intercultural integration and, hopefully, a rallying point for more systematic and broader based efforts in the future.

Although, obviously, titles are only very tentative at the present stage, I would like to call the first volume of my "Principles" something like Humanistic Discussion of Knowledge. As hinted above, my intention here is to present a philosophical theory of knowledge, helpful for the self-understanding of man in the age of science. Accordingly, I plan to divide this book into two parts. In the first part I plan to discuss systematically knowledge as apprehended by man in our age. In particular, I shall take into account the psychogenetical dimensions of knowledge, its sociological components, and its historical aspects.

In the second part of the book, I intend to develop a philosophical synthesis which takes into consideration both the insights of the various philosophical schools and the data of science and history which I have examined in the first part. I am convinced that such a synthesis is both possible and fruitful. It is surprising how mutually enriching—really complementary!—are the epistemological attitudes of both science and philosophy once that one takes the pains of trying to understand them from within.

In line with my general approach, the second book of my "Principles" could be called *A Humanistic Discussion of Reality*. This, again, I intend to divide into two main parts. In the first part I plan to examine systematically the various forms of information about reality available to man in the scientific age. Thus I shall analyze the formal dimensions of reality as presented by mathematics and logic; the interactive dimensions of reality, presented by physics, chemistry and biology; finally, the self-determining dimensions of reality, presented by psychology, anthropology, and the social sciences in general.

In the second part of this book I plan to formulate a philosophical synthesis of the humanistic significance of reality as known to man in the age of science. In particular, I shall discuss how reality presents itself to us as the realm of pervasive intelligibility, intimate beauty, universal dynamism striving toward complexity and order.

The third book of any "Principles" should be *A Humanistic Discussion of Creativity*. In this book I plan to discuss the ethical position of man in the age of science. In the first part, I plan to analyze the new perspectives for creativity that science has opened up to modern man both in relation to intellectual and artistic and practical creativity. Through science, we have come to realize that man is essentially active. He is the maker not only of new intelligibility and new harmony, but also the maker of a literally new world.

In the second and third part of this book I plan to formulate a philosophical synthesis of creativity in the age of science. This I shall do by discussing creativity both as an appeal of values to freedom, and as a personal response of man to a sapiential and religious ideal.

At this point it is likely that you, having patiently followed me so far, feel inclined to ask some questions concerning the program outlined.

Consequently I shall try to complete this letter by giving you the reasons that have brought me to the formulation of the program in question.

I admit that the plan outlined is very difficult, and hardly to be tackled by a single man. Yet what I have in mind to do is not a final synthesis, but rather a tentative sketch, a kind of blueprint for scientific humanism. Someone has to take upon himself the risk of providing the leadership in such an important field. This is then what I intend to do. My hope is that this work will offer to well-intentioned scientists and philosophers a concrete rallying point for a mutually fruitful dialogue. Thus, slowly, a more satisfactory synthesis will be brought about.

As for my personal ability to face such a task, this is certainly a big objection. I came to the conclusion that it was my duty to try only after much reflection and discussion. A number of thoughtful people have encouraged me on. Above all, I was moved by the frankly positive attitude of Professor Heisenberg.

Of course, also practical problems have to be overcome in order to bring about this work. The most serious one is the widespread unwillingness of universities to support initiatives of truly interdepartmental type. On the other hand, foundations usually fund only projects sponsored by universities. Yet, to be consistent with my convictions, I thought I had no choice but to engage myself to the best of my possibilities in order to help heal the split of our culture. As a consequence, I have decided to dedicate myself entirely to this project, while welcoming any kind of help that can contribute to make it a success.

I am now at the end. This letter has become much longer than I expected. But I think I had to comply as well as possible with your request for "a general outline for the proposed series" of my books. You said that it was of "exceeding interest" to you.

I gladly take this opportunity for congratulating and thanking you. You and your staff are so kind, prompt and obliging to certainly deserve the most sincere appreciation. May your generous work for scientific humanism be crowned by all the success it so richly deserves!

Truly yours,

Enrico Cantore

c/o Jesuit Writers' Service 106 West 56th Street
New York, NY 10019

Letter 68. From Enrico Cantore to Annemarie Giese (June 27, 1969)

New York, June 27, 1969

Dear Miss Giese:

I am pleased to have the opportunity to address a letter to you personally. First, I would like to send you my greetings and express my appreciation for your help.

In this context, I would like to once again ask for your assistance. When I was a guest at your institute, I promised several people reprints of two of my articles. Recently, when I prepared to fulfill my promise, I was dismayed to discover that I had somehow lost the necessary addresses. Naturally, I thought of you and your well-known helpfulness.

The individuals in question are partly from your institute and partly from the Seewiesen Institute. From your institute, the individuals include Mr. Heinrich, the engineer; Mr. Hopf; and the doctor from Graz (if I recall correctly), in whose office I worked for a week. He is a long-standing collaborator of Professor Heisenberg, though still young. I have forgotten his name. At the time of my visit, he was substituting for Dr. Dürr.

From the Seewiesen group, there is Mr. Bischof, the psychologist, and probably Dr. Mittelstaedt, one of the directors there.

Here is my request: I have sent two envelopes, each containing five reprints, to you. The two articles are "Science and Philosophy" and "Scientific Humanism."[7] Could you please distribute one copy of each article to the three aforementioned gentlemen at your institute? Regarding the two gentlemen from Seewiesen, Mr. Hopf can handle the matter himself. (If you do not have Mr. Hopf's address, Mr. Heinrich can assist you.)

Thank you for this new proof of your kindness and goodness. May the good Lord reward you abundantly!

With best regards and good wishes,
Yours sincerely,

(Enrico Cantore)

---

7. They are probably manuscripts not yet published in the form of peer-review articles. See Cantore, "Scientific Humanism and the University"; "Science and Humanism."

Letter 69. From Werner Heisenberg to Enrico Cantore (July 9, 1969)

July 9, 1969

Dr. Enrico Cantore
627 East 187 Street
Bronx, NY 10458, USA

Dear Dr. Cantore,

Many thanks for your letter. I am enclosing a copy of my speech in Athens and of my article on Wolfgang Pauli's Philosophical Views, the latter in German, because the original article was written in German. I am enclosing a written permission to quote the text from my interviews which you have seen at the American Institute of Physics.

With best wishes Yours

H.

[Attachment: a letter of permission]

Attachment: a letter of permission

Professor W. Heisenberg

American Institute of Physics
–Library–
335 East U5 Street
New York—USA,

Father Enrico Cantore is permitted to quote and to reproduce any passages he needs from my interviews for the History of Quantum Physics.

    H.

Letter 70. From Enrico Cantore to Werner Heisenberg
(July 16, 1969)

New York: July 16, 1969

Dear Professor Heisenberg:

I acknowledge with gratitude the arrival of your last mail. Best thanks for your permission to quote from your Quantum Physics interviews. Sincere thanks also for copies of your two articles: "Wolfgang Pauli's philosophische Auffassungen" and "Natural Law and the Structure of Matter."[8]

Gratefully yours,

(Enrico Cantore)

---

8. See Heisenberg, "Wolfgang Pauli's Philosophical Views"; *Natural Law and the Structure of Matter*.

Letter 71. From Enrico Cantore to Werner Heisenberg (August 11, 1969)

August 11, 1969

Dear Professor Heisenberg:

I am sorry to disturb you again. But I know your great kindness and willingness to help. Thus I feel confident to request your help once again.

After having asked the advice of Dr. Weiner, the Director of the Niels Bohr Library at the American Institute of Physics, I applied to the American Philosophical Society for a grant. I have asked $1,500. The Society, in fact, frequently helps people who need initial support to start a program of important research and writing. The Society answered me that they want the opinion of four experts. I, of course, thought of you immediately. This is the reason of this letter.

You know my present situation. I have found a possibility of having room and board while helping part-time in a parish, and thus have a great amount of time to dedicate to writing. But this does not give me money. Precisely to have the funds that are needed to start my work here, I have requested the fund mentioned.

The sum given above is just about what is needed. The main expenses are to buy some fundamental office/equipment, buy books and also travel a little in the area centered around New York in order to interview scientists interested in the humanistic and philosophical issues of science.

I enclose here a copy of a summary of my project as sent to the America Philosophical Society. I enclose also a form that should be mailed directly to the Society with your opinion.

I am sure that you will write to the Society as soon as possible.

My poor words are not able to express all my gratefulness to you. But certainly the Lord, who has already made you so great in a fully human sense, will certainly help you to become ever greater in return for your generosity. Please, excuse me for these expressions. You know that they manifest my sincere admiration and love for you.

Thankfully yours,

(Enrico Cantore)

Enrico Cantore
627 East 187 Street
Bronx, NY 10458 USA

Attachment: the applicant's project description

NAME OF APPLICANT: Enrico Cantore   Page three

This page is to be used only if additional information is considered necessary.

(Please typewrite on one side only.)

NATURE OF MY PROJECT AND REASONS FOR SUBMITTING
THIS APPLICATION

I have given much thought to the problem of the relations between science and the humanities. Besides long formal studies, I reflected for many years on the methodological approach, had contact with students as a teacher (Gregorian University, 1965/1967; Fordham University, 1967/1968) and discussed the issue with leading scientists and philosophers. My conviction is that the two-culture split is one of the most serious challenges confronting modem man, and that it cannot be met but by a total dedication to the task. As a first contribution, I published a few papers and authored a book. I was encouraged by the very positive reception of these writings on the part of both experts and educated readers. As a consequence, I decided to devote all my energies to the purpose of bringing about a dialogue and understanding between the two cultures.

I think my principal activity should be writing. Accordingly, my plan calls for a two-fold step. In the first place, I intend to explore in depth the questions at issue. In the second place, I plan to formulate myself some positive guidelines for solution. In short, calling this whole field with the name of Scientific Humanism, I intend to write: (a) an Introduction to Scientific Humanism in two volumes; (b) some volumes which I would call Principles of Scientific Humanism. The book I am now beginning to prepare is the first volume of the Introduction. It intends to explore the humanistic significance of science. That is, its aim is to investigate the new insights, perspectives, and problems disclosed by science to man as regards the understanding of himself and his relationships toward reality. This I plan to achieve by considering science as a concrete factor affecting the whole man, both as an individual and as a social being, in his historical and cultural dynamism. The second volume of the Introduction will explore the relevance of philosophy for man in the scientific age. That is, its aim will be to investigate the historical contributions of philosophy to science, and the reactions of philosophy to scientific progress. In particular, through this research I hope to be able to clarify the complex reasons why a satisfactory understanding between science and philosophy is as yet missing. On the basis of the foregoing Introduction, I expect to be able to present myself some positive indications for a solution of the problems explored. My intention therefore is to write a further series of books which may serve as a sort of first draft or position papers to help stimulate a cooperative dialogue between the two cultures. I am confident that such a dialogue will gradually bring about the mutual understanding and integration desired.

These are the reasons for submitting this application. On the one hand I am convinced that an effective contribution toward bridging the present intercultural split requires a full-time dedication to research and writing. On the other hand, experience tells me that it is hardly possible to achieve this aim by being attached to the paid staff of an institution of higher learning. Such institutions, in fact, generally insist either on teaching or on research in very specialized areas. Consequently I decided to opt for the status of free researcher. This I did out of personal concern for the questions at issue, and also in view of my limited needs for life being single. However, my financial situation is tight. In particular, I find it difficult to meet even the moderate expenses entailed by in the initial stage of my program. Thus I submit this application precisely with the hope of obtaining help for this initial stage.

New York, August 11, 1969

Enrico Cantore

Letter 72. From Werner Heisenberg to American Philosophical Society (August 20, 1969)

American Philosophical Society
104 South Fifth Street, Philadelphia, PA 19106

The Committee on Research of the American Philosophical Society would appreciate your opinion in connection with the application of Enrico Cantore for a grant of $1,500.

Please discuss qualifications of applicant for the proposed research, importance of proposed research, and any other pertinent information:

I have known Dr. Cantore for a few years since we had some discussions in my institute on his book *Atomic Order—An Introduction to the Philosophy of Microphysics* and on general aspects of the relation between old humanistic philosophy and modern science. By our conversations and by several letters we exchanged, I got the strong impression that Dr. Cantore takes a profound interest in the problem of a synthesis of the sciences and the humanities; by his thorough education and his very wide knowledge he is certainly well prepared for doing work in this field. To show my confidence in his abilities I should like to mention that I suggested Dr. Cantore for translating my latest book dealing with physical and philosophical aspects.

Professor W. Heisenberg,
Director of Max-Planck-Institute for Physics and Astrophysics,
Munich
August 20, 1969

Letter 73. From Enrico Cantore to Werner Heisenberg (October 28, 1969)

New York, October 28, 1969

Dear Professor Heisenberg:

I have just received the copy of your recent book *Der Teil und das Ganze* that you so gently sent me immediately upon publication. I am most thankful to you for this act of exquisite kindness. Above all, I deeply appreciate the friendly words with which you dedicate this copy to me.[9] Possibly the best way for expressing my gratitude is to say how much significant for all my work is your generous support. I shall keep this book as a concrete souvenir of your goodness.

My present research on a planned book, tentatively entitled *The Humanistic Significance of Science*, is proceeding satisfactorily. I am now clearly in possession of the fundamental ideas. If everything goes well, I hope to be able to finish the manuscript within 1970 or shortly afterwards. It is precisely in connection with this work (the first one in the planned series of five volumes dealing with *Scientific Humanism*) that I found out how much I owe to you. My sojourn at your Institute, my reading your manuscript and having the opportunity of discussing the issues involved with you personally has made me see deeply what is the humanistic significance of science.

While I heartily congratulate you on your magnificent recent book, I sincerely wish that it be widely known and read because it will contribute immensely to the development of a modern, "scientific" humanism.

I have a good piece of news to communicate to you, and for this, too, I want to express my sincere thanks. The American Philosophical Society has accepted my request for funds to pay the special expenses involved in my research project. They have promised to give me $1000. This sum, though smaller than the original amount requested, is nonetheless of considerable help to me. In addition, there is the encouragement of the recognition given to my work, despite the fact that I am still an unknown person and a foreigner.

Finally a word about my first book, *Atomic Order*. It shall be out in December, at the MIT Press. The Press itself is likely to send you a copy of

---

9. Cantore refers to a handwritten dedication on the cover page. However, no copy of this dedication could be located.

it for publicity, but I intend to send you a personal copy myself, as a little expression of my gratitude and respect.

With all best greetings, I am sincerely yours

(Enrico Cantore)

Enrico Cantore
627 East 187 Street
Bronx, NY 10458 USA

Letter 74. From Enrico Cantore to Werner Heisenberg
(January 17, 1970)

New York, January 17, 1970

Dear Professor Heisenberg:

About one week ago, a copy of my book *Atomic Order* should have been mailed to you by the Jesuit Writers' Service. I hope it will reach you safely in a matter of few weeks.

The few words with which I mention you in the Preface of the same book are sincere. They were written at the depth of my crisis at Fordham. They were true at that time, they have been proved even truer afterwards. It is a matter of fact that your frank encouragement and generous assistance have contributed immensely to give me confidence toward a successful development of my work.

To give expression to my sentiments on this occasion I wrote a simple dedication in the copy mailed out to you. These words come from my heart. Besides thankfulness, I would like to emphasize my sense of respect for you. By this I mean to indicate my joy of having come to know concretely the enriching and humanizing nature of science through my personal contact with you. Such an experience of the humanity of science is a continual source of inspiration for my research and writing.

As you probably know, the book I am currently working on bears precisely the tentative title *The Humanistic Significance of Science*. I would like therefore to dedicate this book in print to you. I expect you will not have any objection to such a public act of recognition. The manuscript should be finished in about a year from now.

    With my best greetings.
    I am truly yours,
    (Enrico Cantore)

P.S. While teaching at The New School of Social Research during the first semester, I mentioned your recent book *Der Teil und das Ganze* to my students. The course was on "The Philosophical Significance of Science." The students are all adults. One of them, an artist named Miss Bettina

Brendel,[10] was so interested that she wrote to Germany for a copy of the book and read it with great care. I was very pleased by that reaction, so I willingly gave her your address when she expressed the desire of writing to you. This young lady is quite a thoughtful person, and very open to the humanistic nature of science.

10. Bettina Brendel (1922–2009) was a German-born Los Angeles painter and educator. Brendel studied at Kunstschule Schmilimsky, Hamburg, and Landeshochschule für Bildende Künste, Hamburg, with Erich Hartmann in the 1940s before coming to the United States in 1951. She continued her studies at the University of Southern California from 1955 to 1958, and taught at the University of California, Los Angeles, from 1958 to 1961. Her art was influenced by atomic physics, and in some of her publications she cited Enrico Cantore and Werner Heisenberg. Bettina Brendel donated her papers (1937–2000) to the Archives of American Art, Smithsonian Institution, Washington, DC, where they can be consulted. See Brendel, "Influence."

Letter 75. From Werner Heisenberg to Enrico Cantore
(January 26, 1970)

January 26, 1970

Dr. Enrico Cantore
627 East 187th Street
Bronx, NY 10458 USA

Dear Mr. Cantore.

Thank you very much for your letter and your kind announcement that you plan to dedicate your next book on the humanistic significance of science to me. I am very pleased about this, as I know that we are in complete agreement on all the essential questions in this field, and I thank you sincerely for your kind intention.

I have responded to the letter from Miss Bettina Brendel.[11] You should caution her against transferring the concepts and images of atomic physics too directly into her artistic world. Nevertheless, I am pleased that she shows so much interest in these philosophical problems.

With many warm regards,
Yours sincerely,

H.

---

11. Brendel's correspondence with Werner Heisenberg has been deposited at the Archives of the Werner Heisenberg Institut für Physik at the Max Planck Institut für Physik und Astrophysik in Munich, West Germany. Records show that Brendel wrote a letter to Werner Heisenberg concerning *Der Teil und das Ganze* and *Physics and Philosophy* on January 6, 1970 (III/93/1723/111–14).

Letter 76. From Enrico Cantore to Werner Heisenberg
(April 16, 1970)

Bronx, NY: April 16, 1970

Dear Professor Heisenberg:

Best thanks to you for your letter of January 26. Your words concerning my interpretation of the humanistic significance of science have been most encouraging for me. I hope I shall not disappoint your expectation. Certainly my conversations with you—particularly the month spent at your Institute—have helped me immensely to understand the profound meaning of science. I expect that the manuscript of the book in question will be finished in about one year from now. Before sending it to the publisher I shall contact you again on this matter.

Knowing your generosity in helping projects aimed at intercultural understanding, I submit to your consideration an initiative which may be significant. A group of Italian philosophers edited a few years ago a *Philosophical Encyclopedia*, which last year came out in second edition of six volumes. Meanwhile, a German and a Spanish translation-adaptations are being prepared. This work has found very good reception by the international public, not only by the Italian one.

Personally, I studied this work in order to write a full-length critical study for *The Review of Metaphysics* in this country.[12] I find it worthwhile, especially because of its attempt to overcome the interdisciplinary divisions. Thus I am in favor of having the Encyclopedia translated and adapted also into English.

A group of American philosophers, too, has come to the conclusion that such a translation-adaptation would be a valuable service to this country. The reason is particularly the fact that philosophy, in the English-speaking countries, is too much under the influence of the empiricist-positivist tradition: even the *Encyclopedia of Philosophy*, published in 1967, is still too heavily prejudiced in that direction. Hence these philosophers have decided to request the help of famous personalities to have their opinion about the suitableness of such a translation-adaptation and also their support, in case that the translation-adaptation is considered worthwhile. The reason for needing the support of famous persons are the large expenses involved. Publishers in this country are not likely to risk capital

---

12. See Cantore, "Italian Philosophical Encyclopedia."

without an assurance that the work would fulfill an important role, hence it can expect to sell a large number of copies.

Since English is an international language, the idea arose of requesting the help of personalities famous on the internal scene. Hence my request to you. If you would have time and possibility of examining some documentation concerning this Encyclopedia, please let me know. I would send you some material which could give you an idea of the work. In particular, since probably Italian would present too great a difficulty for you, I have here available for you German translations of a few articles dealing with scientific matters taken from the second Italian edition. (These translations are being used for the German edition.) In addition, I could send you copies of my critical study for *The Review of Metaphysics* and also a short comparison made by me between the Italian and the American encyclopedias dealing with scientific subjects.

With my most sincere appreciation and greetings, hoping that everything is well with you and your family.

Thankfully yours,

(Enrico Cantore)

Enrico Cantore
627 East 187 Street
Bronx, NY 10458 USA

Letter 77. From Werner Heisenberg to Enrico Cantore (May 20, 1970)

May 20, 1970

Prof. W. Heisenberg

Dr. Enrico Cantore
627 East 187th Street
Bronx, NY 10458, USA

Dear Mr. Cantore.

Thank you very much for your letter and the information about the *Philosophical Encyclopedia*, whose translation into English is being planned. I am very pleased that these broader philosophical directions are also gaining dissemination in America, where science is still too positivistically oriented. Nevertheless, I must unfortunately ask you to excuse me from reviewing these plans, as I am already burdened with numerous other commitments and do not wish to take on any new ones. I hope you will not hold this refusal against me.

    With best wishes for your continued work,
    Yours sincerely,

    H.

V.

"I feel profoundly close to you in spirit."

*Enrico Cantore Gaining Traction,
December 1970–December 1972*

THROUGHOUT THIS PERIOD, CANTORE started gaining recognition as a scholar and intellectual figure, moving from being his mentor's student to almost a peer. This part of their letters shows how they had come to respect each other more, with Heisenberg still in his role as a mentor yet recognizing that Cantore was growing in self-assurance and finding his own voice in relating science and humanism. Heisenberg increasingly viewed Cantore as a colleague whose work was worthy of interest and support.

One of the turning points in this relationship was Cantore's translation of Heisenberg's lecture, "The Meaning of Beauty in Exact Natural Science," given at the Bavarian Academy of Fine Arts in Munich on October 9, 1970.[1] This was a professional assignment for Cantore and a profoundly intellectual challenge. In a letter to Heisenberg, Cantore expressed his gratitude for the opportunity and said that the translation was, in some way, a continuation of their dialogue.

During the fall of 1970, Cantore's *Atomic Order*, published by MIT Press in 1969, was in circulation, and review copies were sent to a small number of academic scientists. Responses were few. Cantore confided his disappointment to Heisenberg: "After one year that the book has appeared, I think I cannot count on it any longer as a means for

---

1. See Heisenberg, "Bedeutung."

establishing a living contact with the scientific community. I am convinced that my work will be a failure if I cannot succeed in having a living and continual dialogue with the scientific community because I am sure that I cannot understand the problematic of scientific man—and much less, help to solve it—without the active cooperation of scientists themselves" (December 17, 1970).

His commitment to bridging the "two cultures," science and humanism, was as much in evidence as ever in his prolific writings during these years. His book *Scientific Man: The Humanistic Significance of Science* was nearing the end of the first draft, and Cantore was anxious for Heisenberg's comments. Despite finding himself isolated and without any institutional backing, Cantore persevered in his mission to establish a serious dialogue between scientists and philosophers. The letters exchanged with Heisenberg during this time reveal fear and resolve: on one hand, he sought reassurance from his mentor, while on the other, he was determined to assert his own scholarly identity.

Their 1971 correspondence marks another pivotal shift in their relationship: Heisenberg began treating Cantore much more as a peer, although he continued to offer advice and moral support. This transition was symbolically marked by their in-person meeting in August 1971. In his letter after the visit, Cantore expressed gratitude for the intellectual exchange and the personal warmth and hospitality extended by Heisenberg and his wife, Elisabeth. This meeting only deepened their bond, and Cantore's reflections highlighted Heisenberg's profound personal and professional influence on his life.

It was also this year that Cantore began to attract some attention. On March 10, 1971, he forwarded an essay to *American Scientist* regarding Heisenberg's *Physics and Beyond*, just published in English. This review would be published in the Swiss philosophical journal *Dialectica*. Cantore's essay highlighted the humanistic vocation of science, a theme central to both his and Heisenberg's intellectual pursuits. Heisenberg was equally excited about the essay, saying that he was pleased Cantore had grasped the core of his philosophy: "You have understood more precisely than most others what I intended to convey in my book, and I am very pleased that your article may help make these intentions understandable to a broader audience" (March 19, 1971).

However, success did not come easily for Cantore, who soon faced a hard reality when trying to publish his manuscript. In a very honest letter dated September 21, 1971, Cantore expressed his self-doubt and frustration over the obstacles blocking his path. Fordham University had cut his teaching courses due to budget cuts, and MIT Press decided not to publish *Scientific Man* due to concerns about its marketability.

Heisenberg's reply, dated October 8, 1971, was typically encouraging and practical. He advised Cantore to continue his work and pointed out that public recognition takes patience and perseverance. Although Heisenberg declined Cantore's request to write a preface—citing his policy of declining all such requests to maintain fairness among colleagues—he offered to write a letter of recommendation instead. Though modest, the gesture demonstrated Heisenberg's belief in the value of Cantore's work and his commitment to supporting him within the limits of his professional ethics.

Cantore was confident his book was a significant scholarly contribution. On December 3, 1971, he wrote to Heisenberg: "As you know, it is a big book because the matters are serious and profound. It would not do service to the public if I did not carefully and thoroughly discuss them. From my experience as a teacher, I know many university students desire a work of this kind. In fact, they become more critical of science not because they oppose intellectual progress, but because they perceive science as inhuman and dehumanizing."[2]

As 1972 progressed, Cantore continued to make progress in his academic career. He applied for a Guggenheim Foundation fellowship, listing Heisenberg as a reference. In his confidential report, dated December 1, 1972, Heisenberg praised Cantore's scholarly contributions, particularly his book *Atomic Order: An Introduction to the Philosophy of Microphysics*. Heisenberg pointed out that Cantore was so deeply concerned with the scientific and philosophical problems of modern physics that he could contribute to the global dialogue between scientists and humanists.

By the end of 1972, Cantore was close to securing a publisher for *Scientific Man*, and his activity was already gaining interest in academic circles. The support he received from Heisenberg during this time was crucial in helping him handle the challenges in his career.

---

2. Paraphrased from the original text, which was written in non-standard English, to improve clarity and readability.

Letter 78. From Werner Heisenberg to Enrico Cantore (October 9, 1970)

[Attachment: The manuscript text of a lecture given to the Bavarian Academy of Fine Arts in Munich on October 9, 1970, later published as Heisenberg, Werner. "Die Bedeutung des Schönen in der exakten Naturwissenschaft." *Physikalische Blätter* 27 (1971) 97–107.][3]

---

3. The manuscript was sent to Enrico Cantore for a translation into English. Belser Verlag was publishing a bilingual version of the Lecture. Cantore was the translator of the parallel English translation from German. The text reports the notes and corrections by Werner Heisenberg. See Letter 85 (June 12, 1971).

Letter 79. From Enrico Cantore to Werner Heisenberg (December 17, 1970)

New York: December 17, 1970

Dear Professor Heisenberg:

On occasion of the coming festivities I wish to send you my most sincere greetings and give some information about my present situation which may be of some interest to you.

I am living in Manhattan at the present. This is much more convenient for me than being in the Bronx. I teach part-time at Fordham University at the Lincoln Center. But my main activity is writing. I have now completed well over half of the first draft of the manuscript of *Scientific Man: The Humanistic Significance of Science*.

I hope to have the final draft ready some time in the spring. As far as I can judge personally, the work is progressing satisfactorily.

Since you have been so generous to take a very encouraging interest in my work, I feel confident to speak to you about my biggest worry at the moment. I refer to my sense of isolation which, if it continues too long, is going to make my whole attempt of helping scientific man to fail. I have no backing of an academic institution of importance. (Fordham U., where I teach, is merely a Liberal Arts College.) Thus, I really counted on my first book as a means to put me in direct touch with the scientific community. The MIT Press has been very cooperative. They have sent out well over one hundred copies to people who may have had a personal interest in my work. A letter was enclosed requesting comments. Only four or five answered acknowledging reception of the book. One professor of Princeton U. even invited me to visit him there, but he frankly admitted that he had not read the book. Thus, after one year that the book has appeared, I think I cannot count on it any longer as a means for establishing a living contact with the scientific community.

I am convinced that my work will be a failure if I cannot succeed in having a living and continual dialogue with the scientific community because I am sure that I cannot understand the problematic of scientific man—and much less, help to solve it—without the active cooperation of scientists themselves. Thus, I am now counting very much on the book I am presently writing for attaining this aim of dialogical contact. This book could succeed where the first did not because its appeal should be wider, and because being the second it does not come from a person totally unknown. However, all efforts should be made that it be really successful.

In this connection I dare to request your help again. Since the book deals with the humanistic way of thinking typical of creative scientists, I am not entirely competent to judge whether I have succeeded in my attempt or not. Thus, I would be immensely thankful to you if you could spare some time to read the manuscript and give me your reactions to it. This would give me the assurance that my book is really meaningful and could be used as a basis for the much-desired dialogue with the scientific community. In addition, I would also feel more at ease in dedicating it in print to you, as I wish to do and you kindly accepted. I know that I am requesting you something that you cannot easily give in view of your many commitments. But I also know from grateful experience that you are really ready to help as far as possible.

I must tell you the truth. If the book I am now writing will succeed, this is due to your having invited me to discuss these issues at length two years ago. Precisely now, doing the writing, I realize more and more how much I learned from your conversation. Thus, I am thinking of you daily, and go back again and again to your marvelous *Der Teil und das Ganze*. In the light of it I have come to understand the whole spirit of science very much better. But, in addition to thinking of you, I am glad to say that I am also praying for you daily. Indeed, from your work I have learned very much about the profound relations between science and religion. So I feel a sincere joy when, every day, while mentioning the various categories of persons for whom I offer the Eucharistic Prayer, I pray to God for "all those who seek you with a sincere heart." In the light of these words, I feel profoundly close to you in spirit. May God bless you abundantly. May he give you good health and the possibility of accomplishing all the good you wish to do.

Thankfully yours

(Enrico Cantore)

Enrico Cantore
220 West 98th Street Apt 8N
New York, NY 10025

Letter 80. From Werner Heisenberg to Enrico Cantore (December 21, 1970)

December 21, 1970

Dr. Enrico Cantore
220 West 98th Street, Apt. 8N
New York, NY 10025, USA

Dear Dr. Cantore,

First, I would like to offer you some consolation regarding the public reception of your book. One should never hope to achieve quick results in terms of changing the opinions of many others through books or other public expressions. Even if a book is ultimately very successful in this regard, its influence only becomes apparent very slowly over the years. It also seems that in America, it is not customary to express gratitude for books received. For my book on the unified field theory, which was also sent to many American colleagues, I received only a very few responses. Nevertheless, I believe I can discern, from the way the younger generation discusses topics at physics conferences, that the ideas in this book have since become widely adopted.

As for your new book, I am happy to review the manuscript. However, whether my opinion will significantly influence the book's long-term impact is something I cannot judge.

With my warmest wishes for Christmas and the New Year,
Yours sincerely,

H.

Letter 81. From Enrico Cantore to Werner Heisenberg (March 10, 1971)

Enrico Cantore
220 West 98th Street Apt 8N
New York, NY 10025

New York: March 10, 1971

Dear Professor Heisenberg:

I enclose here copy of an essay which I have just finished and submitted for publication to *American Scientist*.[4]

The immediate reason for writing this essay has been my desire to contribute effectively to the humanistic understanding of science. The appearance of your book in English can do much to this purpose, but unfortunately many people—even among scientists—are not always able to realize its profound importance. This has been my rather disappointing experience both in reading reviews of your book (for instance, in the *New York Times Book Review* and in *Physics Today*)[5] and while mentioning your book to some renowned scientist.

As a consequence of this situation, I thought I could offer something to the general understanding by making a systematic analysis of the contents of your book as a documentation of the humanizing vocation which science can really be.

I have emphasized to the editors of *American Scientist* that the responsibility for the essay is exclusively mine.

Of course, however, I hope that you may agree with my interpretation of your work. I hope I have not misread nor exaggerated its profound message.

The ultimate reason for writing this essay is—I must confess it quite candidly—my intimate wish of expressing to you my great esteem and thankfulness. Probably the following comparison does not apply

---

4. Attached to the letter was a manuscript titled "Science as a Dialogical Humanizing Process: Highlights of a Vocation." This paper was ultimately not published in *American Scientist* but appeared as Cantore, "Science as a Dialogical Humanizing Process."

5. See Morison, "Poker and Mozart"; Bunge, Review of *Der Teil und das Ganze*, 63–64. Other critical reviews: Hufbauer, Review of *Physics and Beyond*; Morrison, Review of *Physics and Beyond*; Forman, "Historiographic Doubts."

entirely, because circumstances were different in part. But when I read and meditated on your testimonies about the influence of Bohr in your life I could not help thinking of your influence in mine.[6] Certainly it was only after having had the opportunity of discussing at length with you that I came to perceive in fulness the richness and profundity of what I like to call scientific humanism.

I am now about to begin composing the second draft of my *Scientific Man: The Humanistic Significance of Science*. I hope to have the text ready to be mailed out at the end of May. I shall send a copy to you as soon as possible. During August I plan to be in Europe. I would be most thankful if you could give me an interview. I would like to hear your opinion about my writing. For my part I would also wish to submit to you my future plans including some possibility for future collaboration.

I close now with my best greetings. I think I am sincere when I say that I feel very close to you in the tension after ultimates and values—as you expressed it so well in your last book.

Truly yours,

(Enrico Cantore)

P.S. From the enclosed essay you can see, at the end, that I am not entirely pleased with the translation of *Physics and Beyond*. I do not consider this situation too serious for a first edition. But I would certainly wish that a final, better edition be brought out. I suppose that this can be possible. So far I have not mentioned the matter to anyone, including Dr. Anshen. But if you are interested in the matter, and desire my collaboration, I am ready. From what I have heard repeatedly said to me about my writings, my English is truly good enough, at least in helping a translator.

[Attachment: a manuscript entitled "Science as a Dialogical Humanizing Process: Highlights of a Vocation," twenty-eight pages, plus two pages of notes].

---

6. Cantore is referring to a passage in Heisenberg, *Physics and Beyond*, 38–42, which is also discussed in his own "Science as a Dialogical Humanizing Process," 297–98.

Letter 82. From Werner Heisenberg to Enrico Cantore (March 19, 1971)

March 19, 1971

Dr. Enrico Cantore
220 West 98th Street, Apt. 8N
New York, NY 10025, USA

Dear Mr. Cantore,

Thank you very much for your letter and the essay "Science as a Dialogical Humanizing Process: Highlights of a Vocation," which I read with great joy. You have understood more precisely than most others what I intended to convey in my book, and I am very pleased that your article may help make these intentions understandable to a broader audience. Overall, I find the reviews in *The New York Times* and *Physics Today* quite encouraging, as I initially feared that the political chapters of my book might lead some readers to reject it. But in this regard as well, your article can be very helpful.

Once again, my warmest thanks,

Yours sincerely,

H.

Letter 83. From Enrico Cantore to Werner Heisenberg (March 26, 1971)

New York: March 26, 1971

Dear Professor Heisenberg,

I have just received your kind letter of March 19.

I thank you very much for it, I am certainly very glad concerning your opinion of my article. It is a great encouragement for me that I am on the right track while trying to understand and explain the humanistic significance of science.

Precisely in connection with this central concern of my life I take the liberty of writing again so soon. It concerns my planned visit to Europe during the next summer.

My main reason for coming, as of now, is frankly the wish of obtaining a personal interview with you. I have several matters of importance, theoretical and practical, which I would like very much to discuss with you.

The best time for me would be in the last third of August. By that time, in fact, you would have probably an opportunity of taking a look also at the book which I am now composing and hope to finish by the beginning of June.

Personally, to save money, I have found a post as a chaplain on a passenger ship which arrives at Naples at the end of July. It would be marvelous for me if I could use this opportunity. Thus, after my visit to friends and relatives in Italy, I could come to see you and depart from Germany for the States.

Please, let me know as soon as possible whether you think you can give me an interview for the time indicated (as for the specific day, of course, I would write again later). For I need to give confirmation to the shipping line in order to make sure that I will get the post of chaplain.

Best thanks to you for all your kindness.
Truly yours,

(Enrico Cantore)

Enrico Cantore
220 West 98th Street Apt 8N
New York, NY 10025

Letter 84. From Werner Heisenberg to Enrico Cantore (April 2, 1971)

April 2, 1971

Prof. W. Heisenberg

Dr. Enrico Cantore
220 West 98th Street, Apt. 8N
New York, NY 10025, USA

Dear Mr. Cantore,

Thank you very much for your letter. At the end of August, I will probably be at my country house in Urfeld, near Munich, although I have not made any firm plans yet. Perhaps we could correspond again at the end of July regarding your plans and mine.

    With best regards,
    Yours sincerely,

    H.

Letter 85. From Enrico Cantore to Werner Heisenberg (June 12, 1971)

New York: June 12, 1971

Dear Professor Heisenberg:

I enclose here copy of my translation of your lecture "Die Bedeutung des Schönen in der exakten Naturwissenschaft" as received from Belser Verlag at the end of April.[7] I could not make the translation immediately, owing to my writing of the final chapters of my book.

As regards this translation, I have tried to be as literal as possible. Sometimes it is not possible to use always the same word in English when the same word occurs in German, but I have tried to do that as far as I thought it allowable. The only liberty that I took was that of breaking some long sentences into shorter ones, since long sentences sound difficult to an English reader.

There are only two minor points to which I would like to call your attention. Galileo's experiments at the Leaning Tower of Pisa apparently have been proved spurious by modern historians of science. Kepler did indeed find his famous laws by studying experimental data, but the data themselves had been gathered not by him, but by Tycho Brahe. Accordingly, I have put the two passages of yours dealing with this points within square brackets (on p. 8) which I respectfully suggest could be left out of the printed text (both English and German). At any rate, this is my suggestion.

I am sending to you copy of the translation so that you may give your opinion of it. The original is being sent to Belser Verlag. To make sure that my English is satisfactory, I asked an American trained in letters to check it carefully. I am very thankful to you for having had the opportunity of studying this text of yours in depth. I have enjoyed it very much. Through it I had the impression of continuing and deepening a personal dialogue with you. I agree entirely on the basic points. On some other aspects there will probably be time to converse during my next visit in August.

On June 3 I sent you by air a copy of my manuscript *Scientific Man: The Humanistic Significance of Science*. I think you should have received it already by now. I shall be very interested in hearing your comments about it when coming to see you. I hope I have not disappointed your expectations.

---

7. Cantore is getting back on this topic to Heisenberg several months later.

I shall leave this address on July 12. The ship Queen Anna Maria is scheduled to reach Naples on July 28. For the moment it is not yet possible for me to determine with absolute precision when I shall be free to come to Munich or neighborhood. Since I know your many occupations I would like to take as little of your time when visiting you as possible. On the other hand, however, I fear that just one regular interview will not suffice to discuss with you some very important questions, theoretical and practical, about which I have great need to hear your opinion. Therefore, please, let me know what you think about the best procedure to follow. I shall certainly be free and ready to come to the place you will indicate to me (Munich or the neighborhood of the city) by the 25th of August. So, approximately between that date and the end of the month it would be a good time for me. I could live for a few days in Munich and see you when you have some free time. But the problem is that those days are taken up largely by a weekend.

If you can still write to me here, I thank you. Otherwise you can reach me at the beginning of August by writing: c/o Rev. Renato Guidotti, Strada di Superga 70, 10132 Torino-Sassi, Italien. Please add: attendere il suo arrivo.

There are many things I am unable to say well in writing, beginning with the fact that if my manuscript book is going to be successful, as I hope, this will be due very largely to you. Without having known you personally and discussed with you, the book would never have been written the way it was. May the Lord bless you abundantly! With cordial greetings,

Yours

(Enrico Cantore)

Enrico Cantore
220 West 98th Street Apt 8N
New York, NY 10025

[Attachment: a manuscript in English titled Werner Heisenberg, "The Meaning of Beauty in Exact Natural Science" (translated by Enrico Cantore), eighteen pages].[8]

8. A transcription of this text is available at https://inters.org/heisenberg-beauty-natural-science.

Letter 86. From Werner Heisenberg to Enrico Cantore
(June 21, 1971)

June 21, 1971

Dr. Enrico Cantore
220 West 98th Street, Apt. 8N
New York, NY 10025, USA

Dear Mr. Cantore,

Thank you very much for your letter and the translation of my academy lecture. Today, just a brief response in haste, regarding your comment on Galileo's falling body experiments. This critique has also been raised from other quarters, and I have now modified the text in the German publication as follows: "Schon die berühmten Fallversuche, die Galilei wahrscheinlich doch nicht am schiefen Turm zu Pisa ausgeführt hatte, zeigen das aufs deutlichste. Galilei beginnt mit sorgfältigen Beobachtungen ohne Rücksicht auf die Autorität des Aristoteles."[9] Perhaps you can make a corresponding change in the English text.

    I look forward to your visit to Germany and am certain that we can arrange a meeting then.

With best wishes,
Yours sincerely,

H.

---

9. "The famous falling experiments, which Galileo probably did not carry out on the Leaning Tower of Pisa, show this very clearly. Galileo begins with careful observations without regard to the authority of Aristotle."

Letter 87. From Enrico Cantore to Werner Heisenberg (June 27, 1971)

New York: June 27, 1971

Dear Professor Heisenberg:

Best thanks for your letter of the 21. Here is my translation of your modified statement which replaces the first sentence in the manuscript H/C 8.

> Already the famous investigations of the fall—which Galilei did probably not carry out at the Leaning Tower of Pisa—indicate that most clearly.

The remainder, as you have indicated, continues to be as before.

I am certainly very glad at the perspective [sic] of meeting you again and having the possibility of discussing with some leisure matters of common interest.

I shall be grateful if you can let me know soon when and where our meeting will take place. For, if I have to write for hospitality to the Jesuits of Munich, I must do that as soon as possible, since the number of places available is limited.

My first stop in Italy, in the very last days of July, will be Castello de Geronimo, 80069 Vico Equense (Napoli). Later on, until I come to see you, my address will be: c/o P. Renato Guidotti, Strada di Superga 70, 10132 Torino-Sassi. In both cases it may be better add the remark: ATTENDERE IL SUO ARRIVO.

My best greetings and my most sincere wishes,
Truly yours,

(Enrico Cantore)

P. S. I have not transmitted the above correction to Belser Verlag. If you are agreed with it, I hope it will not be difficult have your secretary communicate it to the Verlag.

As regards correction of the proofs, the Verlag did not let me hear anything so far. I shall certainly be available for that work in case my help is needed and an arrangement can be made for a suitable time and place to do it.

Letter 88. From Enrico Cantore to Werner Heisenberg
(August 8, 1971)

Turin: August 8, 1971

Dear Professor Heisenberg:

I have received shortly a letter from Belser Verlag saying that you were agreed with my translation of your lecture. I was really glad at hearing that.

Up to this moment I have not yet received information concerning the place where I will be able to meet you. I have therefore decided to reach Munich in the afternoon of August 25.

My address there is the Jesuit Provincial residence, i.e.,

c/o Ignatiushaus
8 München 22
Kaulbachstraße 31a

By August 20 I can be reached at

c/o Aloisianum
21013 Gallarate (Varese) Italy

Needless to say, I am looking forward to this meeting with you with great expectation and joy. Please let me know when and where I can come to see you. I am ready to travel outside Munich if you are in your summer residence. As you know, I hope to be able to discuss several important matters with you. With best greetings and the hope to seeing you soon again.

I am thankfully yours

(Enrico Cantore)

P.S. I expect to sojourn in Germany (Munich or neighborhood) until the end of the month.

Letter 89. From Werner Heisenberg to Enrico Cantore (August 12, 1971)

August 12, 1971

Dr. Enrico Cantore
c/o Aloisianum
I 21013 Gallarate (Varese)
Italy

Dear Mr. Cantore,

Thank you very much for your letter. I kindly ask you to simply call my office after your arrival in Munich to arrange an appointment for our discussion. I am not yet certain whether I will still be in Munich or already in Urfeld by August 25, but my office will be able to inform you. I look forward to seeing you again.

    With best regards,
    Yours sincerely,

    H.

Letter 90. From Enrico Cantore to Annemarie Giese (August 29, 1971)

D-8000 München 23
Seestrasse 14

August 29, 1971

Dear Ms. Giese

I am writing here the Heisenberg Documents, as agreed. I would like to introduce you to Father Balczewski, a Polish Jesuit. Perhaps you can help him when he comes to visit the Institute. He is also a philosopher and physicist, so he would like to visit the MPI für Physik. Thank you very much for your kind help. May the Lord bless it.

    Yours

    (Enrico Cantore)

Letter 91. From Enrico Cantore to Werner Heisenberg (September 21, 1971)

New York: September 21, 1971

Dear Professor Heisenberg:

The first motive for writing this letter is deeply felt gratitude. Our recent encounter has been for me a greatly enriching experience. The privilege of spending some hours at your home with you and Mrs. Elisabeth has made me realize more concretely how science can acquire a thoroughly human significance, including family life. This experience was for me very heartening. I thank you and your wife for it.[10]

I also wish to express my gratitude for lending me the dossier containing the reactions to your recent book.[11] I was deeply touched by your exquisite kindness in allowing me to peruse your personal documentation.

The second motive for this letter is to inform you of my present situation. Owing to the widespread economic difficulties, Fordham made some budget cuts and dropped my courses. The MIT Press has financial problems, too. Hence their judgment: "extremely complimentary regarding the care with which you prepared the manuscript [*Scientific Man*]" and "very sympathetic to your overall approach"—but the decision was not to publish the book. The reason: "the market for such a book would be very restricted."

I bother you with these personal details only because I know of your active interest in my work. The double blow mentioned made me doubt for a moment whether it would be reasonable at all for me to continue in this almost impossible undertaking. But your repeated assurances in the past gave me new courage. Hence I decided not to abandon the course begun. An unexpected invitation makes it possible for me to continue living in this country despite the present lack of any source of income. A group of teaching brothers in Puertorican Harlem wished to have priest residing with them, while giving room and board. I accepted the invitation and am now trying to develop my program as formerly planned.

---

10. For details on Heisenberg's relationship with his wife, see Heisenberg and Heisenberg, *My Dear Li*.

11. It is likely a collection of book reviews from magazines and academic journals.

I wish to hear your opinion in relation to several points touching upon the continuation of my work. In the first place, the question concerning the possibility of finding sufficient readership for my books. It is obvious that a publisher may be reluctant in risking his money. But, would it be right to say that there is not a market for such books? The interest seems to be widespread in the public. Several recent works of the kind have even become best-sellers. However, meeting the public interest in the matter is not enough. Books may be influential for the wrong motive and thus cause even immense damage: for instance, Monod's *Hasard* and Skinner's *Walden Two* and *Beyond Freedom and Dignity* (just appeared).[12]

If this is the case, it seems to me that one can only do a worthwhile work if one has the courage and the patience of studying in depth and systematically the issues involved. For only in this way can one contribute to the formulation of a better worldview, genuinely humanistic and truly scientific. As you know, this has been my deep motivation and plan in dedicating myself to this work.

My crucial question at this point is: Have I personally such a significant contribution to make so as to justify a continual effort to develop my program despite the almost unconquerable difficulties that confront it? This is an objective question. I would greatly like to hear your opinion about it because, being so deeply involved in this matter, it is only too easy for me to be misled by illusions. But then I should certainly not spend the rest of my life pursuing chimaeras.

If it is justified and even dutiful for me to continue in the course begun, here are the practical goals I envision. The immediate goal, of course, is the publication of SM. But mere publication does not appear sufficient to me. For I can certainly not succeed in my efforts toward a humanistic synthesis unless I acquire some sort of public recognition. That is, I need to gain some respect from the scientific community so

---

12. Monod interprets the processes of evolution to demonstrate that life results purely from natural processes, driven by "pure chance." The basic tenet of this book is that systems in nature, such as enzymatic biofeedback loops, can be explained without invoking final causality. See Monod, *Chance and Necessity*. *Walden Two* is a utopian novel by behavioral psychologist B. F. Skinner, which controversially rejects free will, including the idea that human behavior is governed by a non-corporeal entity such as a spirit or soul. See Skinner, *Walden Two*. In *Beyond Freedom and Dignity*, Skinner argues that belief in free will and the moral autonomy of the individual (what he referred to as "dignity") hinders the use of scientific methods to modify behavior, which he posits as essential for creating a happier and better-organized society. See Skinner, *Beyond Freedom and Dignity*.

as to be able to develop a dialogue with them. Furthermore I need to be recognized somehow by the studying youth so as to exert some influence on them and thus contribute to a better future of our civilization.

But, of course, the attainment of the goals listed is for me extremely difficult for obvious reasons. Publishers do not want to risk their money for a practically unknown author. Scientists tend to dismiss as meaningless the writings of a philosopher. University students can hardly be reached by writings which do not enjoy some sort of popularity. The impasse seems unbreakable—even more so at present that I have lost my academic position as a professor.

I am certain that you are willing to help me break the impasse mentioned with all the means at your disposal. This conviction gives me courage to go on outlining various possibilities that may make the attainment of the goals listed a reality.

To achieve publication of SM a strong letter of recommendation would perhaps be sufficient. But to break the wall of silence and obscurity that surrounds me it seems that something more would be needed. I am thinking of some form of public support or endorsement. What? As we were speaking recently, one such form would probably be that of allowing me to appear in public with you—for instance, on a lecture tour of this country, on a televised interview or the like. I certainly wish very much that something of the kind may come true some day in the future. But the unknowns attending such a form of support are clearly many. For it will depend on the availability of your time, the state of your health, the disposal of money to cover the expenses and—as far as I am personally involved—the previous publication of SM. For, of course, my presentation to the public would be fruitful only if I have something substantial to offer to the public itself.

At least theoretically speaking it would seem that the most effective form of support at the present would be one which would simultaneously most likely obtain the publication of SM and present me to the public. But I feel so terribly ill at ease in mentioning it. For there is nothing in the world I abhor more than offending you. And yet, the apparent desperateness of my situation and your well-known generosity give me courage. I refer here to the idea of your writing a preface for my book. I know of your general rule of not writing prefaces, and I find the rule quite wise. Only, as mentioned, the hopelessness of my situation makes me bold to revert to this topic. One of my readers—Dr. Wilhelm Magnus (mathematics; formerly Göttingen, currently New York University)—feels convinced that a preface of yours would make a complete difference concerning both the publication and the success of my book. This, of course, is also my conviction—both because of your great name as a scientist and the current widespread resonance of your admirable *Der Teil und das Ganze*. Thus I have dared to mention also this possible form of support. But may it suffice to have discreetly submitted it to your consideration. I know your heart from experience and have a great respect for you. I shall welcome whatever decision you make.

I am looking forward to hearing of my possible contribution to the English translation of your philosophical books. If feasible, I shall be most glad to help.

My most sincere thanks to you and to Mrs. Elisabeth. May the Lord continue to bless abundantly you both and your entire family, I pray for the complete recovery of your health.[13]

Sincerely yours,

(Enrico Cantore)

Enrico Cantore
125 East 103rd Street
New York, NY 10029

P.S. If possible, I would be greatly pleased to receive a copy of your new philosophical book as soon as available.[14]

---

13. It is probably during this visit that Cantore becomes aware of Heisenberg's precarious state of health. One of his biographers, Cassidy, explains that Heisenberg state of health rapidly declined after 1969 (now we can explain his reluctance for travels in the US) and a liver condition caused increasing weakness, dizziness, and depression, which eventually required hospitalization. On December 31, 1970, Heisenberg resigned the directorship of Max Planck Institute for Physics (he became director emeritus). "Exploratory surgery indicated advanced cancer of the kidneys and gall bladder for which little could be done. Chemotherapy helped delay the inevitable. In 1975 his condition worsened. He was hospitalized again and returned home too weakened to recuperate." Cassidy, *Uncertainty*, 544–45. On October 1975, he also resigned the presidency of Alexander von Humboldt Foundation.

14. Cantore is likely referring to Heisenberg, *Schritte über Grenzen*.

Letter 92. From Werner Heisenberg to Enrico Cantore
(October 8, 1971)

8 October 1971

Dr. Enrico Cantore
125 East 103rd Street
New York, NY 10029, USA

Dear Mr. Cantore,

Thank you very much for your letter in which you ask for advice—a request that I find quite difficult to fulfill. You know that I have always appreciated your books. Not only do I think the approach you advocate is fundamentally correct, but I also find that you present important questions with an impressive level of expertise, clarity, and accessibility, guiding readers' thinking in a direction that, in my view, belongs to the future. But now comes the question of how to ensure that the book reaches its audience: that a publisher prints it and that it is distributed. The path you propose—that I write a foreword for the book—is unfortunately not an option for me. I would risk severely offending several colleagues to whom I recently declined the same request, and that is something I cannot do. I could, of course, write a recommendation to the publisher with whom you are in negotiations, and I would gladly do so. However, that may not be sufficient, as the publisher must take on considerable risk with such an extensive work. I myself am not in a position to judge when a book will resonate strongly with the public. For example, I cannot understand why Monod's book—which I philosophically disapprove of, as far as I am familiar with it—has become such a commercial success. Or perhaps, as you suggest, it is precisely because his book aligns far too closely with a widespread trivial philosophy (positivism and rationalism)?

You also ask me for personal advice: Should you continue on your current path despite all resistance, or should you scale back or change your goals? You rightly write that you need a certain degree of public recognition to pursue the task you have set for yourself. Public recognition, however, can only be earned slowly, over the years; even a single good book is usually not enough. Perhaps you should focus even more than before, as far as possible, on participating in public intellectual institutions—e.g., as a university lecturer, a participant in conferences and discussions, and so forth. In this regard, you seem to have had some misfortune so far—through no fault of your own, as it appears to me. But it should be possible for you to establish a position at an American university. You would need to take the first steps yourself, as I have virtually no connections left to American academia. However, I would gladly assist with a letter in your favor once the initial connections are made. I want to encourage you to take such steps because I genuinely care about your ideas reaching broader audiences and because I believe you have something valuable to contribute to the intellectual development of our time. From such a base within academia, you would likely find it easier to approach suitable publishers.

I hope you are not too disappointed by this advice, but I cannot offer a better one.

With many warm regards,
Yours sincerely,

H.

Letter 93. From Enrico Cantore to Werner Heisenberg (December 3, 1971)

December 3, 1971

Dear Professor Heisenberg:

I thank you for your sympathetic letter of October 8. I am grateful for the encouragement you gave me. As for the practical ways of making my philosophical contribution to science successful, I must obviously allow circumstances to guide me. But I want to take advantage of all possible opportunities for obtaining an academic position.

At the moment my central concern is the publication of *Scientific Man*. I agree with you that the future should belong to a more humanistic interpretation of science, but I am disturbed by a number of recent facts which seem to point in the opposite direction. You probably know of Born's disconsolate remarks about the negative influence exerted by science on ethics as expressed in his autobiography *My Life and Views*, 1968.[15] The huge success of two recent books seems to prove him right. Skinner's *Beyond Freedom and Dignity*, ever since its appearance about two months ago, has been one of the ten most sold books in this country. The same publisher has now brought out Monod's *Chance and Necessity* and is advertising it heavily.

I am convinced that my *SM* has an important contribution to offer. As you know, it is a large book because the issues involved as are serious and complex. It would not serve the public well were I not to discuss them systematically and in depth. As for finding readers for it, I know from my experience as a teacher that many university students desire a work of the kind. In fact, they become more and more critical of science not because they are against intellectual progress but because they have the impression that science is inhuman and dishumanizing.

Besides the students, of course, numerous reflective scientists and open-minded philosophers expect a clarification on a subject which concerns them much and about which they feel unable to arrive at a satisfactory interpretation through their personal efforts alone.

I gladly accept your kind offer concerning a letter of presentation for *SM*. Since it is quite possible that I shall have to approach more than one publisher, I would suggest that the same procedure be followed as

---

15. See Born, *My Life and Views*.

in the case of *Atomic Order*. That is, I request you to mail your letter to me so that I may be able to present it to the persons who could favor the publication.

Through the present letter I wish to participate personally in the celebrations for your seventieth birthday. So far it has not been possible to me to give a public proof of the great esteem and gratitude I have for you. Pretty soon, however, my study originated by the appearance of your *Physics and Beyond* should appear in the Swiss journal *Dialectica*.[16] During the course of the year I also hope to publish *SM* which will be dedicated to you.

My heartfelt regards and wishes to you. Regards and wishes also to your wife.

Sincerely yours,

(Enrico Cantore)

Enrico Cantore
125 East 103rd Street
New York, NY 10029

P.S. I enclose here copy of my review of *Physics and Beyond* which appeared in September.[17]

---

16. See Cantore, "Science as Dialogical Humanizing Process."
17. Enclosed is a copy of Cantore, Review of *Physics and Beyond*.

Letter 94. From Werner Heisenberg to Enrico Cantore (January 25, 1972)

On the book: 'Man and Science' by Enrico Cantore

One of the most important problems of our present time concerns the relation between the so-called 'two cultures,' i.e., between science and its consequences on the one side, the 'humanistic' aspects of life on the other; or, to put it simply, between science and philosophy. In his book 'Man and Science' E. Cantore tries to give a thorough analysis of the many different aspects of this relation. There is the humanistic relevance of science, which had been ignored for a long time, then the necessity of philosophical studies for the understanding of modern science, the responsibility of the scientist for the consequences of his research and many other similar problems. E. Cantore can base his investigations on a thorough knowledge of traditional philosophy and its history, and he is well acquainted with several branches of modern science, especially with atomic and nuclear physics, and with their epistemological aspects. He is convinced that the commonplace mixture of pragmatism and positivism is not sufficient for dealing with the present situation, and he tries to elucidate the difficulties of our time by invoking more general philosophical concepts and by considering the historical roots of modern science. I could imagine that the book could help many, especially young people to find their way through the intellectual confusion of the present world, to get a clearer view of the merits and the limitations of modern science and technology, and to connect this knowledge with a new understanding of the humanistic values. Therefore, I would like to recommend the publication of the book in every way.

Munich, January 25, 1972

Prof. Dr. W. Heisenberg

[Handwritten annotation:]

Send to Dr. Cantore with thanks for the congratulations.

Letter 95. From Enrico Cantore to Werner Heisenberg
(March 29, 1972)

Enrico Cantore
125 East 103rd Street
New York, NY 10029

New York: March 29, 1972

Dear Professor Heisenberg:

Please excuse me for not having acknowledged so far your most encouraging recommendation of my *Scientific Man*. I have delayed because I wished to give you the good news of the acceptation of the manuscript for publication which always seemed imminent. After several difficulties, I am practically sure to have found the publisher. I have just signed the preliminary contract with Learned Publications, which is the American branch of Nauwelaerts of Louvain-Paris.

Other good news. My study on your *Der Teil* has appeared in *Dialectica*.[18] It will be used as the basis of a seminar that I will give at Columbia's Institute for the study of Science in Human Affairs on April 24.

I enclose here copy of an article of mine.[19] I wrote it at the invitation of Dr. Yourgrau who heard of me from you early in 1970.[20] It was not deemed suitable for the type of journal that *Foundations of Physics* wanted to be. But Margenau, the co-director of the journal, recommended it for publication to *Philosophy of Science*. Despite its being one of the directors of the latter journal, there was considerable reluctance in accepting it. In the end, however, the result was most encouraging. For about one month now I keep receiving requests for reprints from scientists active all over the world. So far eighty of them have written for a copy. (But no

---

18. Cantore, "Science as Dialogical Humanizing Process."

19. Cantore encloses a copy of Cantore, "Humanistic Significance of Science."

20. Wolfgang H. J. Yourgrau (1908–1979) served as professor of history and philosophy of science at the University of Denver, Colorado. He was associated with the University from 1963 to 1978. He also served as chair of the Philosophy Department and was a member of the History Department. He earned his PhD in physics from Humboldt University, Berlin, Germany in 1932. During World War II, Yourgrau edited *The Orient*, an anti-fascist, German-language weekly. He was editor of *Foundations of Physics*, an international periodical he founded with Henry Margenau, Yale physicist and spectroscopy expert.

philosopher has written to me.) I thought you would enjoy seeing it, since you gave the first inspiration that led to its actual composition.

There are several additional developments I wish to bring to your attention. In the first place Fordham in Manhattan invited me to teach there regularly, but only every second year in order to save money for their budget. In this connection they told me that they were applying to a foundation for money in order to invite an outstanding scientist, interested in humanistic issues, for a public discussion of these issues with their students. They suggested that I informed you about this initiative, for they would like very much to have you. Of course, for the moment there is no need of any commitment on your part, since the money is not yet available. But if you could come it would be marvelous, as I will explain further at once. As for the time, the early date for which the money should be available could be the spring of 1973, but a more likely time will be the early autumn of 1973.

The specific reason why I would be most thankful for your coming here would be the invaluable help you would give me in my effort to reach and serve the scientific community. I have seen from the reaction of so many scientists to my article that I really can give something worthwhile to them. But the opposition of philosophers—especially that of the Neo-positivists who dominate the field of philosophy of science—is also mounting. One book review of my *Atomic Order* could not have been more spitefully destructive.[21] Also the great reluctance of *Philosophy of Science* against the enclosed article must have stemmed from the same source. As a consequence, to succeed in my efforts of dialogue and service I need to gain such a respect from the scientific community that it cannot be destroyed by the attacks of the positivists. To this end, therefore, if you could come for a visit, your help would be of immense significance. For if I could appear with you in public—in an interview, for instance—scientists around this area would recognize that my efforts deserve consideration, and thus would more easily be willing to enter a dialogue with me. Then the systematic effort of dialogue and exchange with the scientific community I have in mind could be developed and bring fruits to make science an effective means for the self-humanization of contemporary man. I have a concrete plan for

---

21. Cantore is referring to Forman, Review of *Atomic Order*. Other reviews of the book had been generally positive: Binns, Review of *Atomic Order*; D'Agostino, Review of *Atomic Order*; Dialectica, Review of *Atomic Order*; Kane, Review of *Atomic Order*; Russo, Review of *Atomic Order*; Whitbeck, Review of *Atomic Order*.

a cooperative undertaking which I have already discussed with several people who have given their approval of principle. If you wish, I shall be glad to write in a more detailed way about it to you.

If you can come here, such an undertaking could be launched on occasion of your visit. For the reasons given, the time I would like most for your possible visit would be not before the early autumn 1973. The reason is also that, by that time, my book *Scientific Man* should be finally published and available for distribution. I know you will certainly do whatever you can to support this initiative. Hence I thank you most sincerely in advance.

A final word. Learned Publication is preparing a cooperative work on contemporary man and the problem of God. If you could write an article on God and science from your experience it would be most welcome. The article should be finished for approximately the next autumn. I really hope that your health has been good despite the rigors of winter.

The best paschal blessings on you, your wife and all your family.

Sincerely yours,

(Enrico Cantore)

Letter 96. From Enrico Cantore to Werner Heisenberg
(September 29, 1972)

New York: September 29, 1972

Dear Professor Heisenberg:

At the suggestion of friends interested in my work I have just submitted an application to the Guggenheim Foundation. Since you have always sponsored my activity so generously, I took the liberty of giving your name as one of the four experts to whom the Foundation will write for an assessment of my project. I thought it advisable to inform you at once about the contents of my application. Hence the enclosed papers.

My news are good. I am teaching at Fordham, on alternate years, part-time. As regards the publication of *Scientific Man*, everything indicates that it should begin pretty soon. A committee of businessmen is ready to supply the money.

Also concerning the dialogue between science and the humanities, there are some hopeful signs. I have already met several scholars who wish to cooperate in the establishment of a center for dialogue.

One of the most promising members of the center for dialogue is Loren R. Graham, a scientific graduate and Ph.D. in history of science.[22] His current position is Associate Professor of History at Columbia University. He has already authored two books, the second of which is rated quite highly by specialists (*Science and Philosophy in the Soviet Union*; Knopf, 1972). Dr. Graham, who is now thirty-nine, has decided to study systematically the history of science and philosophy in the twentieth century. Hence our hope of working together. Hence also the reason for mentioning him to you. He expects to be visiting Europe during this

---

22. Loren R. Graham (born 1933) is an American historian of science specializing in the history of science in Russia. He is currently professor emeritus of the history of science at Harvard University. Loren R. Graham received his BS in chemical engineering from Purdue University in 1955 and his PhD in history from Columbia University in 1964. He served as professor of history at Columbia University from 1972 to 1978, before becoming professor of the history of science at the MIT. Graham has been awarded fellowships from the Woodrow Wilson Foundation, the Danforth Foundation, the Guggenheim Foundation, and the Rockefeller Foundation. His academic work focuses on the history of science and the study of contemporary science and technology in Russia. He is a fellow of the American Association for the Advancement of Science (AAAS) and the American Academy of Arts and Sciences, a member of the American Philosophical Society, and a foreign member of the Russian Academy of Natural Science.

school year. In the course of his visit, he would request—if possible—a personal interview with you. I promised to him that I would recommend him warmly to your attention.

I do not know whether you still think of coming to the United States in the foreseeable future. In case that this would be the plan, please inform me as soon as possible. Many people, of course, would like very much to hear a public lecture of yours. As regards the financing of your trip, it should not be a problem—especially now that I have found several wealthy businessmen earnestly interested in the human aspects of science.

I wish you the best health and much inward consolation in your work and in the thriving of your family. My warmest regards also to your charming and lovable wife. I remember the evening I spent at your home as one of the finest days of my life.

Sincerely yours,

(Enrico Cantore)

***

NARRATIVE ACCOUNT

The central goal of my life is to contribute to the solution of the so-called problem of the two cultures. I was led to this goal by a series of steps as follows.

My training has been in three different areas: philosophy, physics and theology. This training brought to my attention the complex nature of the cultural problem of modem man.

I first decided to explore the philosophical significance of science.

To this end I developed an approach which I called inductive-genetic and applied, as a test case, to the investigation of atomics physics. I intended to find out what are the philosophical presuppositions and implications of concrete scientific research. I discovered that, by doing science, man becomes a culturally new person: he thinks and acts in a way which is quite different from that of his nonscientific counterpart. I published the results of my investigation of atomic physics in my book *Atomic Order: An Introduction to the Philosophy of Microphysics* (MIT

Press, 1969), and decided to pursue systematically the insight concerning the all-encompassing cultural significance of science.

I settled in the United States because this is the leading scientific country of the world, its language is the universal means of communication among scientifically interested people, there is here a widespread interest in the meaning of science for man. Despite the well-known difficulties of interdepartmental cooperation inside universities and the financial troubles of the nation during the past few years, experience has proved that my decision was justified. The following is the research work I have carried out up to the present.

I discussed in a number of essays the methodology for a comprehensive understanding of science from the humanistic standpoint. These essays appeared in several journals. The reaction of experts has been favorable. In particular, the response to my synthetical article in Philosophy of Science has been worldwide. So far 142 scientists, belonging to all branches of research and living in all six continents, have written to me for reprints. (I enclose copy here for documentation, as a summary of my approach.)

I examined systematically the new perspectives and problems affecting the comprehension of man which arise when the whole of science is taken into account. This study is finished in manuscript form, under the title *Scientific Man: The Humanistic Significance of Science*. The readers of this manuscript have all been quite positive in their evaluation of its scholarship and exhaustiveness. Only, its scholarly character and bulk present difficulties for publication, owing to the current financial straits of university presses. A committee of New York businessmen is presently collecting the money needed to subsidize the printing. To make the results of my inquiry more universally accessible, I have also written a college-level presentation of the same subject entitled *Science and Dignity: The Challenge to Man*. This popularization will appear as soon as my scholarly work will have been published.

At the present I am researching another major book aimed at complementing the analysis undertaken in the preceding work. There I considered mainly the standpoint of the scientist. Here I take particularly into account the standpoint of the philosopher. Philosophers, over the centuries, have repeatedly attempted to formulate a comprehensive doctrine of man. As is well known, they have not yet succeeded satisfactorily in their endeavor. The book I am now researching investigates the historical and cultural reasons for such a failure and the lessons to be learned for the achievement of a comprehensive humanism which really suits the scientific age. I submit this application because I need help to complete this investigation.

Postdoctoral Grants and Fellowships I have received

Gregorian University: investigation of philosophical aspects of quantum physics and acquaintance with current trends in history and philosophy of science in the English speaking world. Amount: traveling and living expenses for stay at the University of Santa Clara, Calif. (1962/1963) and University of Chicago (1963/1964).

Professor Jean Piaget: invitation to participate as a member in his Symposium on Genetic Psychology, University of Geneva (one week, June 1967). Amount: traveling and living expenses.

Fordham University: organization of a center for dialogue between science and humanism (1967/1968). Amount: 6,000 dollars.

Professor Werner Heisenberg: visiting fellowship at his Max-Planck-Institut für Physik, Munich to discuss with him the humanistic significance of science. Amount: services (office, secretary, etc.). Stay: October 1968.

American Philosophical Society: support of research for Scientific Man. Amount: $1,000 granted on October 16, 1969.

New York, September 26, 1972

(Enrico Cantore)

PUBLICATIONS

Articles

"Philosophy in Atomic Physics: Complementarity"
  Modern Schoolman 34 (1957), pp. 79–104.
"La Soluzione Fisica dell'Enigma Quantistico"
  Divus Thomas (Plac.) 60 (1957), pp. 150–59.
"La Sapienza Biblica, Ideale Religioso del Credente"
  Rivista Biblica 8 (1960), pp. 1–9; pp. 129–43; pp. 195–205.
"Genetical Understanding of Science: Some Considerations About Optics"
  Archives Internationales d'Histoire des Sciences 19 (1966), pp. 335–563.
"Science and Philosophy: Some Reflections on Man's Unending Quest for Understanding"
  Dialectica 22 (1968), pp. 132–66.
"Scientific Humanism and the University"
  Thought 43 (1968), pp. 409–28.
"The Italian Philosophical Encyclopedia"
  The Review of Metaphysics 24 (1971), pp. 510–52.

"Science and Humanism: The Sapiential Role of Philosophy"
   Dialectica 24 (1970), pp. 215–41.
"Humanistic Significance of Science: Some Methodological Considerations"
   Philosophy of Science 38 (1971), pp. 595–612.
"Science as Dialogical Humanizing Process: Highlights of a Vocation"
   Dialectica 25 (1971), pp. 293–316.
"Science and Religion: An Experiential Analysis"
   Extensive essay (eighty pages) contributed to S. A. Matczak, ed. *God in Contemporary Thought* to be published in 1973 by Learned Publications, New York and Editions Nauwelaerts; Louvain University Press.[23]

Books

Atomic Order: An Introduction to the Philosophy of Microphysics
   (Cambridge, MA: MIT Press, 1969)
Scientific Man: The Humanistic Significance of Science
   To be published in 1973 by Learned Publications, New York and Editions Nauwelaerts; Louvain University Press.
Science and Dignity: The Challenge to Man
   Negotiations are being conducted at the present for its publication.

Translation

W. Heisenberg, *Die Bedeutung des Schönen in der exakten Naturvissenschaft*
   Published in bilingual edition by Belser Verlag, Stuttgart, 1971.

Requested book reviews

J. A. Ripley, *The Elements and Structure of the Physical Sciences* (1964), in *Gregorianum* 47 (1966), pp. 617–19.
G. F. McLean, ed. *Proceedings of the American Catholic Philosophical Association*, vol. xxxviii; History and Philosophy of Science (1964), in *Gregorianum*, pp. 619–20.
S. Dockx and P. Bernays, eds. *Information and Prediction in Science* (1965), in *Gregorianum* 48 (1967), pp. 412f.
E. Cantore, *Atomic Order: An Introduction to the Philosophy of Microphysics* (1969), in *Atti della Fondazione Giorgio Ronchi* 25 (1970), pp. 681–89.
W. Heisenberg, *Physics and Beyond; Encounters and Conversations* (1971), in *Theological Studies* 52 (1971), pp. 516–18.

New York, September 26, 1972
(Enrico Cantore)

---

23. This essay was not published in this book (1977), and it was not possible to find a draft.

STATEMENT OF PLANS

My application for a Fellowship covering the 1973/1974 academic year is motivated by the following reason. The overall goal of my work is to contribute to the solution of the so-called two-culture problem. The continual development of science and its technological applications call into question the system of certainties and values which enabled prescientific man to have a sense of cultural identity. Accordingly, to face the cultural problem of scientific man satisfactorily, one must examine it from several complementary angles. I have hitherto endeavored to explore the issue from the standpoints of the scientist and the philosopher. I wish now to consider also the cultural impact of the scientific mentality on man as a social being. Since I am not a sociologist and since the evaluation of the data available in this area presupposes an interdisciplinary approach, it is hardly possible for me to obtain satisfactory results by simply studying published material. Hence my wish to travel in order to consult experts in the field. This is my reason to apply for a Fellowship.

I plan to dedicate one full year free from teaching duties to the investigation mentioned. My approach consists of two successive stages. In the first place, I intend to contact as many experts as possible in prominent institutions both in this country and abroad. Personal acquaintance will enable me to decide about the few most suitable institutions where to study the issues in depth. In the second place, I plan to carry out intensive research in the institutions which will prove to have most to offer for my purpose.

The Centers of interest to my plan are the following:

(1) The Hudson Institute: Croton-on-Hudson, NY

(2) The Center for the Study of Democratic Institutions: Santa Barbara, Calif.

(3) Science Policy Research Unit, University of Sussex, Great Britain

(4) Science Studies Unit, University of Edinburgh, Great Britain

(5) Group 2000: Amersfart, Holland

(6) UNESCO, Paris

(7) The Club of Rome, Rome

(8) TRADES, Rome

(9) Centre d'Études Prospectives—Association Gaston Berger: Paris

(10) Max-Planck-Institute for the Study of Life in the Scientific-Technological World (director: C. F. von Weizsäcker): Starnberg, Bavaria

(11) Max-Planck-Institute for Ethology (director: K. Lorenz): Seewiesen, Bavaria.

On the basis of the information currently available to me the institutions which are most likely to offer me a profitable program for research in residence are the three last ones in the list above. (9) is one of the earliest and internationally best regarded centers aimed at developing a comprehensive philosophy of man in the scientific age. (10) tackles systematically the problem of man, especially the pressures that come to bear on him in highly developed scientific and technological nations. (11), by studying evolution ethologically, promises to shed important light to illuminate man's own evolution as a social being.

My sole source of income is part-time teaching. In addition, I have currently room and board as a resident chaplain. (The religious order to which I belong—Society of Jesus—allows me free use of my time, but gives me no financial support whatever). Since the research outlined is such as to preclude the possibility of earning any money during one full year, I request the Foundation to give me funds to cover both my living expenses and extensive travels during the same time. After consultation with some experts, I think that I would need $10,000.

I sincerely thank the Foundation for taking my request into consideration.

(Enrico Cantore)
New York, September 26, 1972

Letter 97. From Werner Heisenberg to Enrico Cantore (October 1972?)

Dear Sir:

May I recommend to you the book *Atomic Order—An Introduction to the Philosophy of Microphysics* by Dr. Enrico Cantore. I have studied the book rather carefully and I find it a very good representation of the philosophical aspects of modern atomic physics. I think that for many students who have learned quantum theory only as a practical tool for solving problems of atomic physics, this book will give a first and very thorough insight into the important philosophical problems connected with quantum theory. The fact that quantum theory actually contains the first mathematical representation of what one may call autonomous totality is taken as being the central point in the philosophical interpretation of modern physics. The recognition of this fact could for many students open up the way to a much deeper and richer understanding of the philosophical side of the problem with which he had to do in his practical work. The representation of these ideas in the book seems to me very readable and understandable. Therefore, I hope that the book will find a very widespread attention and interest among the younger generation of scientists.

    Sincerely yours

    H.

Letter 98. From Guggenheim Memorial Foundation to Werner Heisenberg (November 21, 1972)

John Simon Guggenheim Memorial Foundation
90 Park Avenue • New York, New York 10016

November 21, 1972

Dear Mr. Heisenberg:

This note brings you our annual request for assistance. We shall be grateful for your candid and critical appraisal of the candidate who has referred us to you. You may be sure that your statement will be held in strictest confidence.

With the Foundation's thanks as well as my own.

Sincerely yours,

(James F. Mathias)
Administrative Vice President and Secretary

Letter 99. From Werner Heisenberg to Guggenheim Memorial Foundation (December 1, 1972)

John Simon Guggenheim Memorial Foundation—90 Park Av., New York, NY 10016
Confidential Report on Candidate for Fellowship
Requested of:
Dr. Werner Heisenberg
MPI für Physik
8 München 23 Föhringer Ring 6

Name of Candidate: Cantore, Enrico

My acquaintance with Dr. Enrico Cantore was initiated by his book *Atomic Order—an Introduction to the Philosophy of Microphysics*. I have discussed the content of this book with him, and I had the impression that Dr. Cantore takes a profound interest in the problems of the synthesis of the sciences and the humanities and that he has a thorough knowledge in both fields. Perhaps I should simply add the comment which I have once written on his book "Man and Science":

"One of the most important problems of our present time concerns the relation between the so-called 'two cultures,' i.e., between science and its consequences on the one side, the 'humanistic' aspects of life on the other; or, to put it simply, between science and philosophy. In his book 'Man and Science' E. Cantore tries to give a thorough analysis of the many different aspects of this relation. There is the humanistic relevance of science, which had been ignored for a long time, then the necessity of philosophical studies for the understanding of modern science, the responsibility of the scientist for the consequences of his research and many other similar problems. E. Cantore can base his investigations on a thorough knowledge of traditional philosophy and its history, and he is well acquainted with several branches of modern science, especially with atomic and nuclear physics, and with their epistemological aspects. He is convinced that the commonplace mixture of pragmatism and positivism is not sufficient for dealing with the present situation, and he tries to elucidate the difficulties of our time by invoking more general philosophical concepts and by considering the historical roots of modern science. I could imagine that the book could help many, especially young people to find their way through the intellectual confusion of the present world, to get a clearer view of the merits and the limitations of modern science and technology, and to connect this knowledge with a new understanding of the humanistic values. Therefore, I would like to recommend the publication of the book in every way."

The application for a fellowship of Dr. Cantore hence seems to me well founded, and I can only approve of his plans. Therefore, I can support his application in every way.

Signed: gez. W. Heisenberg
Director (em.) Max-Planck-Institute for Physics and Astrophysics
Föhringer Ring 6, D-8000 Munich 40 (FR of Germany)

Date: December 1, 1972

VI.

"I wish to convey to you some encouraging news."

Final Letters and Positive Steps Forward for Scientific Humanism, August 1973–June 1976

THE CORRESPONDENCE BETWEEN ENRICO Cantore and Werner Heisenberg ended in June 1976, just as important developments were taking shape in the realization of Cantore's vision of scientific humanism. From August 1973 to June 1976, the Institute for Scientific Humanism was founded, and Cantore's long-term efforts to publish *Scientific Man: The Humanistic Significance of Science* were finally successful. These letters reflect an unbroken hope despite the challenges Cantore faced and show his unwavering dedication to advancing the integration of science and humanism. Heisenberg passed away peacefully at his home on February 1, 1976.

Founding the Institute for Scientific Humanism

Cantore's letter of August 30, 1973, announced the founding of the Institute for Scientific Humanism at Fordham University's Lowenstein Center. For Cantore, this was quite the capstone: he had dreamed for years about an institute exactly of this sort. The ISH was to tackle the intellectual crisis of the present time by creating interdisciplinary dialogue and promoting a comprehensive image of man in the scientific age. It was intended to bridge the sciences with the humanities—an endeavor long central to Cantore's work and one that Heisenberg strongly supported.

Cantore's initial definition of the ISH outlined its objectives, organizational form, and research subjects it wanted to involve. He especially underlined the independence, interdisciplinarity, and dialogic nature with which members would approach their work. Naturally, the ISH was also to foster contact, research, and teaching, first and foremost through reports, directories, and interdisciplinary seminars—high aspirations that Cantore considered realizable since he believed science combined with humanistic values could bring a more complete understanding of the human condition.

Heisenberg's Continued Support and Cantore's Struggles

Throughout 1974, Cantore continued to inform Heisenberg about the progress of the ISH, seeking advice and support from his mentor. In a letter from January 1974, he informed Heisenberg about establishing the ISH, underlining the philosophical and interdisciplinary nature of the institute's mission. Heisenberg was very encouraging and appreciative of ISH goals that might inspire young generations to look at science in more humanistic ways. His approval greatly encouraged Cantore, who valued Heisenberg's insights and support.

But hard facts of funding and institutional support held back Cantore's optimism. He described his attempts to secure funding for the ISH in letters he wrote in 1974. Cantore had applied to several foundations for grants, but he was generally frustrated by the philosophical bent of the ISH's work; it was too abstract and speculative and, therefore, not practical enough for funding bodies that favored projects based on demonstrable achievement. Cantore's frustration is palpable in a June 14, 1974, letter, where he pours out his disappointment at the lack of progress and having to put personal savings into the ISH to keep it afloat. During such trying times, continued support from Heisenberg was invaluable.

In his response of February 20, 1974, Heisenberg's last letter to Cantore, the German scientist was relatively encouraging—once more stressing the ISH's mission and potential to solve the "problem" of the human factor in scientific progress: "I would hope that your new institute allows the younger generation to participate in fostering the use of science for building a more human world." This letter, which Cantore later attached to funding applications, underscored the long-range significance of Heisenberg's mentorship and his belief in Cantore's efforts.

The Publication of *Scientific Man* and Heisenberg's Legacy

Although financial problems remained during this period, Cantore never relented in his academic work. By June 1976, he had gathered enough contributions to fulfill his long-cherished desire to publish *Scientific Man*, which he dedicated to his mentor and friend Werner Heisenberg. In his final letters, Cantore reflected on the profound impact of Heisenberg's thought, mainly his focus on the creative and humanistic aspects of scientific inquiry.

Heisenberg's death in February 1976 only strengthened Cantore's resolve to continue the work they had discussed over the years. In a letter dated June 15, 1976, Cantore evoked Heisenberg's "responsiveness," a central element of his scientific creativity and humanism, legacy, and ISH guiding principle.

Letter 100. From Enrico Cantore to Werner Heisenberg
(August 30, 1973)

Institute for Scientific Humanism
Lowenstein Center at Fordham University
60th Street and Columbus Avenue
New York, NY 10023
Enrico Cantore
Director

August 30, 1973

Dear Professor Heisenberg:[1]

I am about to leave New York for an extensive visit in Europe. I wish to convey to you some encouraging news.

As you see from this letterhead we have finally begun what had remained a dead project during my first stay at Fordham in 1967/1968. Many difficulties remain to be overcome, but the outlook is promising.

Another good piece of news concerns my manuscript book *Scientific Man* which you kindly recommended for publication. All the technical and financial preparations are now practically settled. Typesetting should begin in the very near future.

I plan to be in Munich from the 11th to the 20th of October. During this time I would appreciate very much the opportunity of discussing with you the plan of ISH. To this end I enclose here copy of the general ideas that have guided me and my friends here in organizing the Institute. This document is not yet printed because I want to obtain observations and suggestions before making it final. Of course, I would welcome your suggestions very greatly.

In Munich I shall stay with my confreres. Dr. Walter Kerber, 8 Munich 22, Kaulbachstrasse 33 (tel. 0811) 28 60 77 will be my host.[2] Please,

---

1. Cantore writes this letter almost a year after the last one (Letter 96, September 29, 1972). In the meantime, at Fordham had been founded the Institute for Scientific Humanism, the center desired both by Cantore and Heisenberg. The German Nobelist in the same year would go to the US and deliver a Lecture at Harvard University on the historical development of the concepts of quantum theory, but there is no record of a visit of Heisenberg at the ISH, nor does Cantore report about one in other letters. See Heisenberg, "Development of Concepts."

2. Walter Kerber (1926–2006), SJ, was a German social ethicist and Catholic theologian. In 1967, Kerber was appointed professor of ethics and social sciences at the

let your secretary send me a note to inform me about the general time you may be available. Then I shall phone for a specific appointment, if possible.³

I hope you and your wife are in good health. I was glad to hear some time ago from Kerber about your Guardini Award.⁴ I also read your speech. This event seems to me very encouraging. I congratulate you sincerely. Best greetings to you and your wife.

Truly yours,

(Enrico Cantore)

P.S. My Deputy Director will mail you a copy of his book on interviews he had with prominent scientists recently on the human and religious implications of science.⁵ Probably you will find it interesting.

[Attachment 1: a description of the ISH, tagged: ISH 8/1973]
[Attachment 2: a thorough description of how the ISH operates, dated April 24, 1973, seven pages]

Jesuit-sponsored Berchmanskolleg in Pullach in the Isartal. After moving to Munich in 1971 under the new name Hochschule für Philosophie (HFPH), he continued his teaching there until his retirement in 1994. From 1976 to 1986, he was also head of the Institute for Social Policy of the HFPH. In addition, he headed the research and study project of the Rottendorf Foundation "Questions of a New World Culture" at the HFPH from 1984 to 1995. In addition to teaching the philosophical subjects "general ethics" and "social ethics," his primary interest was Catholic social doctrine. Kerber died as a result of long-standing Parkinson's disease.

3. There is no further correspondence concerning this meeting in Munich around October 11–20, 1973. From Letter 102 (February 8, 1974) we learn that the two met in Munich during Summer 1973.

4. Werner Heisenberg was awarded the Guardini Prize on March 23, 1973. The Romano Guardini prize is an award given by the Catholic Academy of Bavaria. This was the first time that the prize was conferred to a non-theologian. An English translation of his speech was later published as Heisenberg, "Scientific Truth and Religious Truth."

5. Cantore is referring to Trinklein, *God of Science*.

Attachment 1: a description of the ISH, tagged: ISH 8/1973

Institute For Scientific Humanism

An Autonomous Center for Interdisciplinary Dialogue
in the Scientific Age

Lowenstein Center at Fordham University
60th Street and Columbus Avenue
New York, NY 10023

General Description

The contemporary intellectual is likely to experience a profound crisis, for his conscientious dedication to learning often fails to supply him with an authentically human message of ultimate significance.

Science is a pivotal factor of the intellectual crisis. Science upsets man deeply because it continually seems to question his most cherished certainties and values.

This intellectual crisis necessitates the rethinking of man's self-understanding and his entire worldview. In particular, an effort must be made to synthesize the new perspectives disclosed by science with the perennial insights supplied by the humanities.

The difficulty of attaining a humanistic synthesis in our times is paradoxically one of abundance. The plentifulness of data tends to make the individual thinker feel isolated and discouraged because of his apparent inability to master all the insights which are relevant to the understanding of man. The Institute for Scientific Humanism is unique in its attempt to meet the humanistic challenge through a corporate effort.

The Institute arose from the spontaneous convergence of humanistic interests manifested by an international group of scientists, philosophers and men of letters. It intends to serve in a mediatory capacity by transmitting information and stimulating reflection.

The principal asset of the Institute is the earnest concern for man which is characteristic of intellectual leaders. The Institute
  requests their contributions of experience and wisdom in order to make them available to the public, and especially to students.

The Staff

THE DIRECTOR. *Dr. Enrico Cantore* is a philosopher with training in physics and theology. His chief interest is the exploration of the significance of science for the whole man. His main contribution to date has been methodological. Besides numerous essays, he has authored three books: *Atomic Order; An Introduction to the Philosophy of Microphysics* (1969); *Scientific Man; The Humanistic Significance of Science* (in press); *Science and Dignity; The Challenge to Man* (in manuscript). He has taught at the Gregorian University in Rome and is currently Adjunct Associate Professor of Science and Humanism at Fordham University/Lincoln Center, New York City.

THE DEPUTY DIRECTOR. *Mr. Frederick E. Trinklein* is an educator and writer with special training in physics and astronomy. He has co-authored several successful science textbooks in current use. His *The God of Science* (1971) contains the interviews he conducted with thirty-eight leading scientists in ten countries regarding the human and religious meanings of science. He is Dean of Faculty and Science Chairman at Long Island Lutheran High School. He also teaches astronomy at Long Island University and Nassau Community College.

THE TREASURER AND FINANCIAL CHAIRMAN. *Mr. Eugene P. Foley* is an investment businessman with an education in philosophy and law. For several years he held senior positions in the United States Government. His writings include a book on black economic development entitled *The Achieving Ghetto* (1968). He is a member of the National Bureau of Economic Research.

THE SECRETARY. *Mrs. Margaret Trinklein* has taught on the primary and secondary levels. She has held secretarial positions in executive office and educational institutions. She is presently editorial assistant for her husband's scientific textbooks and other publications.

Attachment 2: a thorough description of how the ISH operates, dated April 24, 1973, seven pages

Institute For Scientific Humanism

Autonomous Center for Interdisciplinary Dialogue
on Man in the Scientific Age

1. Definition

ISH is an autonomous organization aimed at promoting the dialogical understanding of man in the scientific age. *Humanism* is taken here as synonymous with the comprehensive doctrine of man.

The *autonomy* of ISH is both *institutional* and *ideological*. ISH is not part of any institution, academic or otherwise. Its unique commitment is to a *consistent open-mindedness*. That is, it intends to study man without excluding any of the approaches that have been traditionally considered necessary to this end. They are, besides science: art, history, philosophy, religion.

2. Origin

ISH has arisen as a result of the personal experience of its founders. After long study of the significance of science for man and of the perspectives that arise therefrom, they have become convinced of the necessity as well as feasibility of an original organization in this area.

The *necessity* of ISH is proved by the fact that, in the present time of specialization, we cannot adequately understand ourselves except by means of *dialogue*. But institutions already existing are insufficient for this purpose. Universities are too large and split along departmental lines. Institutes dealing with science and man ordinarily concentrate on practical issues such as ecology, politics, economics, ethics. None—at least in the English-speaking world—seems to exist which tries to get to the theoretical roots of the current humanistic crisis in a sufficiently broad and systematic manner.

The *feasibility* of ISH is proved by the positive response from many scientists and other scholars to the humanistic writings of its founders.

3. Goal

The goal of ISH is *theoretical*. It consists in working toward a comprehensive vision of the whole man living in the age of science. The main avenues ISH has chosen to pursue its goal are communication, research, and education.

Communication: humanism demands a synthesis which no individual can bring about alone. ISH exists to mediate an exchange of ideas an information among those who feel a need for it.

Research: humanism presupposes a systematic investigation of many areas. ISH exists to foster such a research of its members and others.

Education: humanism aims at helping man form himself. ISH exists to provide the basis for a better formation of man, especially at the college level.

4. Structure

ISH is a corporate undertaking. *Membership* is open to all professionals and interested laymen who share in the dialogical spirit and are willing to contribute to the achievement of the common goal according to individual possibilities.

The *leadership* comprises: the President; the Director with the Executive Staff made up of Deputy Director, Secretary, Treasurer and Associates; the Advisory Committee [see appended Profile]; the Financial Committee.

ISH is run by the Director with the assistance of the Staff, in consultation with the President and the Chairmen of the two Committees.

Organized research projects are undertaken by Institute *Fellows*.

5. Program

The initial activities programmed by ISH are the following:

Newsletter: a periodical intended in the first place to supply information about current publications and events of interest to scientific humanism. Further services will be considered according to possibilities and suggestions from the readers.

Directory: a project aimed at covering systematically the contemporary scene in scientific humanism. We intend to list periodicals, publication series, organizations, persons, and the like.

Interdisciplinary Seminar: an experimental undertaking aimed at developing practical guidelines for a satisfactory dialogue among specialists in different disciplines.

Open Lecture Series: an initiative intended to popularize the humanistic significance of science among the studying youth. Its originality consists in a unified-experiential approach. Each set of lectures is to deal with a single theme. The selected speakers are expected to comment on the theme by reflecting on their own personal experience as creative scientists and humanists. Proceedings will constitute the Library on the Human Meaning of Science.

Further activities will include conferences of experts on topics of special current interest, workshops for professionals, philosophical investigations.

All members of ISH are invited to present suggestions and contribute information, especially in order to insure the usefulness of the Newsletter and the Directory.

6. COLLABORATION

ISH intends to collaborate as widely as possible with individuals and organizations working for the welfare of man in the scientific age. In particular, it wishes to place its theoretical resources at the disposal of:

Institutions of Learning: by suggesting approaches, programs, literature, lecturers, etc.

Ethical and Social Agencies: by trying to clarify the philosophical issues which underly their specific concerns.

As far as possible, ISH also desires to contribute to the illumination of public opinion, but only indirectly.

7. LOCATION

The physical facilities of ISH are located in the Loewenstein Tower on the Fordham University Campus at the Lincoln Center. However, ISH is and intends to remain a completely independent institution. The compensation for the use of the University facilities is regulated by contract between the two administrations.

NOTE. This is the second draft of the ISH plan. I thank for the various comments on the first one (dated March 10, 1973). Please, let me have further reactions and suggestions.

New York, April 29, 1973

Enrico Cantore, Director
125 East 103rd Street
New York, NY 10029
Phone: 427.5928

***

Profile of ISH Advisory Committee Member

The *purpose* of the Advisory Committee is to ensure the intellectual excellence and comprehensiveness of the work undertaken by the Institute for Scientific Humanism. In view of this goal it seems that the necessary and sufficient conditions to be fulfilled by its Members can be reduced to the following general characteristics and concrete engagements.

General Characteristics

The Advisory Committee Member should be:

An established scholar—a person who has contributed, in a significant way, to his field of learning

Working in one of the major areas of interest for a comprehensive understanding of man—the various sciences, the humanities (art, literature, etc.), history, philosophy, theology

Interested in the reflective study of man living in the scientific age—a person who realizes the seriousness of the present crisis of man and is determined to contribute to its theoretical solution

In agreement with the methodological principles of ISH—these principles being its *theoretical goal* (development of a comprehensive vision of contemporary man) and its *dialogical approach* including *all* issues and perspectives affecting man as a whole.

Concrete Engagements

The Advisory Committee Member should feel bound by:

No ideological precondition of any kind (religious, philosophical or otherwise). He may be a religious believer or unbeliever, may belong to any or none of the philosophical schools.

Consistent dialogical open-mindedness relative to the issues affecting the understanding of man in the scientific age. Although perfectly free in his ideological position, he must not exclude the discussion of any topics which are generally regarded of importance for understanding man (these topics being of a scientific, artistic, philosophical, historical and religious nature).

A minimum of organized contribution: accept that his name be publicly linked with ISH; acknowledge written communications from the Staff of ISH; participate in occasional meetings (if and when the majority of the Advisors would agree on the desirability of such an initiative).

Personal availability to inquiries for advice concerning his own field of expertise, when the questions are posed for the direct furtherance of the aim of ISH.

Willingness to offer spontaneous suggestions to the Staff of ISH concerning initiatives to be undertaken, persons to be contacted, bibliographic information to be considered and the like.

NOTE This is a tentative outline. I shall welcome all suggestions which will help to improve it.

New York, April 29, 1973
Enrico Cantore, Director

"I WISH TO CONVEY TO YOU SOME ENCOURAGING NEWS."

Letter 101. From Enrico Cantore to Werner Heisenberg
(January 1974)

Institute for Scientific Humanism
Lowenstein Center at Fordham University
60th Street and Columbus Avenue
New York, NY 10023
Enrico Cantore
Director

January 1974

Dear Colleague and Friend:[6]

Thinking that it may interest you, I wish to announce the recent foundation of our Institute. We are a new autonomous center for interdisciplinary dialogue on man in the scientific age. Here are our motivations, goals and projects.

ISH is a response to a widespread desire we detected within the scientific community. One decisive factor was the reaction to an article of mine which summarizes my philosophical approach to science: "Humanistic Significance of Science: Some Methodological Considerations." Even though it was published in a specialized journal (*Philosophy of Science* 38 [1971], pp. 395–412), it attracted the attention of an unexpectedly large number of scientists. About 150 of them, living in all six continents and working in all branches of research, wrote to me for reprints.

Another decisive factor was the outcome of the interviews that Frederick E. Trinklein conducted with some forty leading scientists in ten countries on the human and religious significance of science (published in his *The God of Science* [Grand Rapids: Eerdmans, 1971]). Even though he was a total stranger to them, they not only welcomed him, but many urged that the dialogue be continued and deepened. In response to these reactions, we joined forces and established ISH. Mr. Trinklein is the Deputy Director.

Our goal is humanistic in the traditional and nonsectarian sense of the term. Humanism in the Renaissance, and all along in the academic usage of the word "humanities," has stood for a comprehensive view of

---

6. This is a circular letter not specifically addressed to Werner Heisenberg; however, Heisenberg was one of the recipients.

man—one aimed not only at understanding but also at inspiring him as a whole. But the current cultural crisis stems clearly from the lack of such a comprehensive view. Man is upset because he no longer knows who he is or what he should strive for. We have founded ISH to meet this humanistic need. In particular, we are convinced that science has an essential role to play in humanism. (For a documentation of this role of science, see, for instance, W. Heisenberg, *Physics and Beyond: Encounters and Conversations*. I have discussed the message of this book in my essay "Science as a Dialogical Humanizing Process: Highlights of a Vocation" in *Dialectica* 25 [1971], pp. 293–316.)

Our approach is philosophical because we aim at understanding man by means of systematic reflection. Many organizations are currently at work exploring the problem of man and science. But largely they concentrate on urgent practical issues or tend to dwell on the logical analysis of scientific discourse. We are trying to complement their efforts by rethinking the conception of man himself, his nature, his values, his role in the world.

The interdisciplinary character of our initiative is obvious from both its prehistory and its goals. Not only did ISH arise from the spontaneous convergence of many interests, but we are dependent on an interdisciplinary approach in order to attain our objectives. In particular, as philosophers, we are well aware that we stand in need of the data of positive research as well as the insights of the traditional humanities.

Our institutional autonomy is a corollary deriving from the goals outlined. Interdisciplinary dialogue is our main approach. But dialogue requires freedom from all constraints that arise from the identification with particular institutions and their ideologies. Thus ISH is and intends to remain an autonomous institution. In particular, though we are grateful to Fordham University for having offered us our initial facilities, our complete independence must be stressed. A contract between the two institutions regulates the use of these facilities.

Our initial project is to spread humanistic information. We are convinced that a principal cause of the contemporary crisis is lack of communication. Intellectuals often become overspecialized because they feel isolated. Accordingly, to foster humanism, we want to begin by supplying information in a twofold manner. First, we shall issue a periodical newsletter to inform our members of recent publications that throw new philosophical light on man in the scientific age. Second, we shall compose a directory on science and humanism in the contemporary world (societies, journals, events, personalities, etc.).

Another major project we are working on is an interdisciplinary forum. One of the main roots of the current crisis is the dissatisfaction of students with either the humanities or science or both. The young are dissatisfied because they do not obtain intellectual leadership. Their elders feel at a loss because suitable interdisciplinary structures are missing. We intend to provide one such structure. We will invite recognized masters to discuss publicly the problem of man in the age of science by reflecting on their own experience as creative thinkers and teachers. Their contributions, organized according to a systematic plan, will then be made widely available by means of a series of books.

To close, I wish to emphasize the cooperative character of our initiative. We are only too aware that, despite our best will, it is practically impossible for us to provide the services listed unless many persons volunteer their cooperation. We are thankful that a number of people have already decided to do so. If you are not one of them, we request that you, too, consider helping us. Contributions can be of various kinds, according to circumstances and opportunities.

To present ourselves more concretely, I enclose a copy of our biographical data and a preliminary list of our Advisors. Additional literature about ISH will be mailed to you as soon as feasible. In return, please let us know whether you are interested in our work and if we have your correct address. We shall appreciate both your suggestions and your questions.

With cordial greetings,

Dr. Enrico Cantore

Attachment. Biographical notes and list of ISH Advisors

Staff of the Institute for Scientific Humanism

The Director Dr. Enrico Cantore is a naturalized American of Italian birth (Turin, 1926), He is a graduate of the State University of Turin (laurea in physics) and the Gregorian University in Rome (doctorate in philosophy). He has also a degree in theology. He conducted postgraduate research in history and philosophy of science at the University of Chicago and the Max-Planck-Institut für Physik, Munich.

Dr. Cantore's central interest is the investigation of science as a factor affecting the understanding of the whole man. His approach is philosophical. His main contribution to date has been methodological. His views have been influenced by numerous encounters and correspondence with such prominent scientist-philosophers as Werner Heisenberg, Vasco Ronchi, Ferdinand Gonseth, and Jean Piaget.

Dr. Cantore's writings include a number of essays and three books. They are *Atomic Order: An Introduction to the Philosophy of Microphysics* (Cambridge, MA: MIT Press, 1969); *Scientific Man: The Humanistic Significance of Science* (New York: in press); *Science and Dignity: The Challenge to Man* (in manuscript).

The reception of Dr. Cantore's views has been favorable among scientists and humanists concerned with the problem of the philosophical understanding of man living in the scientific age. A scattered but worldwide audience is interested in his writings. This situation has moved him to join forces with a group of friends and start the Institute for Scientific Humanism. He is currently organizing the Institute as its Director.

Dr. Cantore is a priest and a member of the Jesuit order. He is also Adjunct Associate Professor of Science and Humanism at Fordham University/ Lincoln Center. He resides in New York City, Address: Institute for Scientific Humanism, Lowenstein Center at Fordham University, 60th Street and Columbus Avenue, New York, NY 10023, Phone: (212) 427-5928.

Deputy Director Prof. Frederick E. Trinklein is an educator and writer with special training in physics and astronomy. Among his writings are *Modern Space Science*, the first American astronomy text for high schools, and *Modern Physics*, which has long been a standard in its field. He also collaborated on *An Introduction to Astronomy*, a college text presently in its second edition. In another book. *The God of Science*, he discusses his interviews with thirty-eight leading American and European scientists regarding the human and religious meanings of science.

Prof. Trinklein is Dean of Faculty and Science Chairman at Long Island Lutheran High School. He teaches astronomy at Nassau Community College and for Long Island University, He has conducted solar eclipse expeditions to Mexico and East Africa and is planning a similar expedition to West Australia for the summer of 1974.

PRELIMINARY LIST OF ISH ADVISORS (JANUARY 1974)

Dr. Hansjochem Autrum—Chairman, Department of Zoology, University of Munich

Dr. Amiya Chakravarty—Professor of Philosophy. State University of New York, New Paltz, NY

Dr. Anthony Chullikal—Philosopher and Economist. International Commission for Justice and Peace, Rome

Dr. René-Marc Habachi—Director, UNESCO Division of Philosophy, Paris

Dr. Peter Edward Hodgson—Senior Research Fellow and Lecturer of Physics. Corpus Christi College, University of Oxford

Dr. J. Allen Hynek—Chairman, Department of Astronomy, Northwestern University, Evanston, Illinois

Dr. Marius Jeuken—Dean, Department of Biology and Director, Institute for Theoretical Biology, University of Leiden, Holland

Dr. Arthur B. Komar—Professor of Physics and Dean, Belfer Graduate School of Science, Yeshiva University, New York City

Dr. Wilhelm Magnus—Professor of Mathematics, Polytechnic Institute of New York, Brooklyn, NY

Dr. Aurelio Peccei—Economist and Founder of the Club of Rome, Rome

Dr. Vasco Ronchi—Director, National Institute of Optics, Florence; former President, International Union of History and Philosophy of Science

Dr. George William Shea—Associate professor of classics and Dean, Liberal Arts College, Fordham University at Lincoln Center, New York City

Letter 102. From Enrico Cantore to Werner Heisenberg (February 8, 1974)

Institute for Scientific Humanism
Lowenstein Center at Fordham University
60th Street and Columbus Avenue
New York, NY 10023
Enrico Cantore
Director

February 8, 1974

Dear Professor Heisenberg:

Probably you have already received our first circular mailed out a few days ago. It gives an idea of how far our organizational effort has progressed to date. It is encouraging to see these initial results achieved amid many difficulties.

You know how much I owe to you in this regard. Thus I am certain that you rejoice with me.

Another encouraging piece of news regards the preliminary contacts we have taken with several foundations in order to obtain operating funds. Their reactions seem to be favorable. In this connection I take the liberty of requesting once more your help.

To obtain a favorable decision from the foundations, it is important that we submit also some opinions from recognized scholars on the significance of our program. My request would be that you put briefly in writing the opinion you so kindly uttered to me last summer when we met in your office. If I remember well, you said approximately that our program deserved being undertaken because it was original and useful and apparently feasible. Please send your statement to me in a letter that I may photocopy and send to the foundations.

I hope to hear from you soon. I also hope and pray that everything is well with you and your wife, best greetings to both. I shall keep you constantly informed of our developments. I am still cherishing the plan of inviting both of you to New York as guests of ISH.

Cordially yours,

(Enrico Cantore)

"I WISH TO CONVEY TO YOU SOME ENCOURAGING NEWS."

Letter 103 from Werner Heisenberg to Enrico Cantore
(February 20, 1974)

Prof. W. Heisenberg

February 20, 1974

Dr. Enrico Cantore
Institute for Scientific Humanism
Lowenstein Center at Fordham University
60th Street and Columbus Avenue
New York, NY 10023, USA

Dear Dr. Cantore:[7]

I thank you for your letter and the information concerning your new Institute for Scientific Humanism. You know that I have always been interested in the human side of science. I had the impression that the disappointment of the young generation from science was largely due to this lack of insight in the human aspect of the scientific process; and as a consequence, in the possibilities to build a more human world by science and by the combination of science and humanities. The program of your institute seems to me very useful and also feasible in this respect. It is a well-known fact that new ideas are not accepted because they are correct or convincing, but because they offer the opportunity to cooperate by one's own activity. In this sense I would hope that your new institute gives an opportunity to the younger generation to participate in fostering the use of science for building a more human world.

With best wishes
Yours,

H.

---

7. This is the last letter Werner Heisenberg wrote to Enrico Cantore. Heisenberg passed away two years later. Although his health deteriorated, he maintained contact with others until late 1975. There is no record of further correspondence, but, as the wise saying goes, "the absence of evidence is not evidence of absence."

Letter 104. From Enrico Cantore to Werner Heisenberg (June 14, 1974)

Institute for Scientific Humanism
Lowenstein Center at Fordham University
60th Street and Columbus Avenue
New York, NY 10023
Enrico Cantore
Director

June 14, 1974

Dear Professor Heisenberg:

I am ashamed that I have waited so long to thank you for your very encouraging letter of support of February 20.

You have expressed the purpose of ISH in a way that has given me much encouragement and stimulus. My delay has been due to the desire of reporting some positive outcome of our applications to numerous foundations. Unfortunately for the moment we have achieved nothing.

The situation of ISH is extremely serious as far as money is involved. I began relying on the promises of some friends. These promises proved to be empty. Thus I had to pay everything out of my savings: more than $3,000 so far.

Also the expectations we had from foundations turned out to be practically empty. We applied to fifteen. Only two took us seriously, the others turned us down without even considering our application. One of the two, Ford, has just turned us down. The other, Rockefeller, had promised what seemed to be a positive answer by April; now they said that they will decide by September.

The situation is grim because of two main reasons. Foundations are public-relations outfits. We learned from experience that one cannot expect any consideration from them unless one has some friends inside. The other reason is the pragmatism prevalent among intellectuals of this country. Thus, when foundations take science and man into account, they support only initiatives with technical features: sociological or psychological surveys, computer uses, new teaching techniques, and the like. A philosophical reflection is simply scorned by them.

It seems to me that I should not give up the effort as long as there is any glimmer of hope. For I consider this undertaking of mine a real service that I must give to contemporary man. Hence, if I would abandon the initiative, I would consider myself guilty of treason.

If you would know some persons in influential positions inside foundations, I would request your help. It would be good if you would write a personal letter to these people presenting me and ISH. If I could have a copy of this letter, I would then take up contact with them. I feel embarrassed to bother you again, but I cannot help it. Possibly your great name could be effective if you write to those who know you personally.

My warmest greetings to you and your wife. I hope that everything is well with both of you. May the Lord reward abundantly your generosity.

Sincerely yours,

(Enrico Cantore)

Letter 105. From Enrico Cantore to Werner Heisenberg
(September 5, 1975)

Institute for Scientific Humanism
Lowenstein Center at Fordham University
60th Street and Columbus Avenue
New York, NY 10023
Frederick E. Trinklein
Deputy Director

September 5, 1975

Dear Colleague and Friend:

You may have been surprised by my long silence following my first circular of January 1974. I wish not to delay any longer reporting to you about our corporate developments that took place in the meantime.

The circular was well received. I want to thank many of you who replied with personal letters offering evaluations, suggestions and other forms of assistance. This correspondence was of great help to us.

The persons who so far have showed an interest in our work are predominantly scientists, some of world renown. They number approximately three hundred. Even though this number is not large, we feel encouraged by it because this group is truly international and interdisciplinary. Furthermore it has come about almost by itself, simply through occasional contacts, without any organized drive.

The corporate initiatives we have undertaken since the foundation of ISH have been organizational.

The Institute was incorporated by the State of New York in April 1974; shortly thereafter it was granted tax exemption by the Internal Revenue Service. At the moment, our Board of Directors includes, besides myself, Professors Frederick E. Trinklein, Deputy Director (Astronomy. State University of New York), Arthur B. Komar (Physics, Yeshiva University) and Wilhelm Magnus (Mathematics, Polytechnic Institute of New York).

Our next concern has been to secure a financial basis for the Institute. In this connection, however, we have been unable so far to make any headway.

We have already contacted some twenty potential sources of support. This disappointing situation accounts for my long silence.

Despite the persisting difficulties of the economy, we are determined to continue our efforts to make ISH financially viable. If we succeed, I shall report to you as soon as possible. Meanwhile be assured of our appreciation for your interest in our endeavor.

With cordial greetings, sincerely yours,

Dr. Enrico Cantore

Letter 106. From Enrico Cantore to Annemarie Giese
(June 15, 1976)

Institute for Scientific Humanism
Lowenstein Center at Fordham University
60th Street and Columbus Avenue
New York, NY 10023
Enrico Cantore
Director

June 15, 1976

Dear Colleague and Friend:

I wish to bring to your attention an event that will probably interest you. Through some donations, our Institute has been able to initiate the production of my book *Scientific Man: The Humanistic Significance of Science*. We hope this publication will mark the beginning of our corporate development. For the book aims to provide a systematic understanding of the humanistic potentialities of science. Hence it should facilitate, for both scientists and humanists, the cooperation required to develop a vision of the whole man that really suits our scientific age. (A brief presentation of the book is given below.)

The book is dedicated to the memory of Werner Heisenberg because, among the numerous scientists I consulted when preparing it, he contributed the most in terms of personal insight, continued interest and encouraging assistance. He knew of my intention to dedicate it to him and graciously agreed. Unfortunately publishing difficulties prevented me from offering it to him as a homage during his lifetime.

The aspect of Heisenberg's humanistic message about science that struck me most in our private conversations has been *responsiveness*. He was in agreement that science is creative; however, in his view, the root of such creativity was nothing but a responsive attitude. Being a gifted amateur musician, he used to resort to a musical simile. The creative scientist—he contended—is nothing better than the ordinary person, except for one respect: he has a more sensitive ear than the average and responds more readily and perseveringly.

This attitude of responsiveness may very well be taken as the key to understanding the humanistic and humanizing import of science.

He seems to have intended as much when, in his eulogy of his friend and colleague Otto Hahn, he emphasized the latter's greatness by writing: "Perhaps his outstanding human and scientific success was deeply rooted in his unconditional 'Yes' to life in spite of all difficulties and his ability to transmit this cheerful 'Yes' to his co-workers and friends," (*Physics Today* 21 [1962], 102).

Given the outstanding character of Heisenberg's gifts and some momentous decisions with which he was confronted, it is necessary to await the results of historical research in order to understand fully his legacy to mankind. However, the responsiveness message will certainly remain a basic component of this legacy. For it expresses one of the central experiential aspects of science as an activity of the whole person. Hence the reason to stress it in connection with our Institute. This message proves that it is not only possible but eminently fruitful to strive for all-encompassing humanization while doing science. In fact, what counts in both the scientific enterprise and the humanizing endeavor is ultimately the same: *responsiveness*, consistent and persevering to the end. In other words, humanism and science clearly belong together and complement each other, to the benefit of man as a whole.

We feel encouraged by Heisenberg's message especially with regard to our intention to make scientific humanism understandable and appealing to the youth. Young people are inclined to respond to the ideal if presented to them properly. This is what we intend to do by taking into account both science and the humanities. We are pleased to say that Heisenberg himself was in full agreement with this intention. Commenting on our program he wrote: "I had the impression that the disappointment of the young generation from science was largely due to this lack of insight in the human aspect of the scientific process; and as a consequence, in the possibilities to build a more human world by science and by the combination of science and humanities. The program of your Institute seems to me very useful and also feasible in this respect" (Letter dated February 20, 1974).

While we offer our heartfelt condolences to Heisenberg's widow, children and friends, we renew our commitment to this programmatic goal.

Cordially yours,

Enrico Cantore, Director

A Presentation of
SCIENTIFIC MAN: THE HUMANISTIC SIGNIFICANCE OF SCIENCE

This is a major scholarly research exploring the contributions of science to the understanding of contemporary man and his cultural world. The approach is experiential-reflective. Science is studied as an experience involving the whole personality of its practitioners, A systematic reflection is undertaken to bring out the structure and originality of this experience and the perspectives that it discloses for an updated philosophy of man which truly suits the needs and expectations of the scientific age.

The approach adopted is developed concretely through an examination of the process of scientific discovery as documented by history, psychology and sociology of science—with particular emphasis on the testimonies of the scientists themselves. The implications of scientific discovery are surveyed methodically to clarify the message about the meaning and purpose of the cosmos accessible to man in the scientific age and the inspiration and challenge that this message offers for a renewal of his ethical values and ideals.

The views of this book have matured over a period of some twenty years, in ongoing contact with the international scientific community. It is addressed to thoughtful scientists, concerned humanists and mature university students both in science and the humanities. Prepublication reactions have been very positive.

The page size of this book is cm. 17 x 25; the number of pages will be approximately 450/500. The price will be $20. It Is scheduled to appear in the Fall of 1976, a production of ISH PUBLICATIONS, New York.

Letter 107. From Enrico Cantore to Annemarie Giese (June 29, 1976)

Institute for Scientific Humanism
Lowenstein Center at Fordham University
60th Street and Columbus Avenue
New York, NY 10023
Enrico Cantore
Director

June 29, 1976

Dear Miss Giese:

Please excuse me for not having written to you before. I learned about the death of Professor Heisenberg from the obituary in the New York Times.[8] On that occasion I wrote my condolences to Mrs. Heisenberg, but it escaped my mind that I had to write to you, too.

I try to make up for this omission with the present letter and its enclosure.[9]

As you know, I am very obliged to you because of your continual cordial help. I hope you will find some consolation for the death of your long-time employer by hearing that the benefit I derived from my contacts with him and you is now about to take a concrete shape in the form of a book. This book is briefly presented in the enclosure.

I have prayed for you repeatedly in the past months. May the Lord give you much peace and serenity in these moments of trial. And may He give you the grace of continuing to do much good in the years to come!

Please, accept my most cordial greetings.

Sincerely yours,

(Enrico Cantore)

---

8. See "Atomic Pioneer," 25.
9. Cantore is probably referring to Letter 106 (June 15, 1976).

Epilogue

THE GREAT PHYSICIST WERNER Heisenberg, one of the men who changed the course of modern science with his breakthrough in quantum mechanics, died at his home on February 1, 1976. During the last years of his weakening health, his loyal secretary, Miss Annemarie Giese, handled much of his correspondence. Among the letters she worked with was the last correspondence between Heisenberg and Enrico Cantore.

The last letter between Heisenberg and Cantore at the Max Planck Institute Archive dates back to June 29, 1976, and it is by Cantore to Giese. It is a thanksgiving letter he wrote right after the news of Heisenberg's death. Cantore shows his respect and admits that Heisenberg's impact on his own work was incalculable. It marks the end of a long correspondence, which has bridged almost one decade and has been extremely enriching for Cantore both professionally and personally.

Heisenberg did not live to see the publication of *Scientific Man: The Humanistic Significance of Science*, which finally appeared in 1977. The front page of the book bore a dedication to Heisenberg, evidence of the Cantore's intellectual debt to the German Nobel laureate: "To the revered memory of Werner Heisenberg, scientist and humanist, with perennial gratefulness." The publication of *Scientific Man* marked a milestone in the career of the Italian philosopher, even if critical reception to the book was decidedly mixed. The book was well received in philosophical and theological circles, and journals such as *American Scientist* and *Leonardo* carried favorable reviews that cited its serious value.[1] However, it had difficulty reaching a wider readership within the scientific community.

1. See Selvaggi, Review of *Scientific Man*; Russo, Review of *Scientific Man*; Shinn, Review of *Scientific Man*; Archie, Review of *Scientific Man*; Landsberg, Review of *Scientific Man*; Clarke, Review of *Scientific Man*.

In an unpublished *memoir* on the occasion of his fiftieth anniversary as a Jesuit, Cantore summed up the three kinds of humanistic-scientific dialogues he practiced along the lines of his research and career: the person-to-person, the organized, and the growth-directed dialogue.[2] Among the many great names he met along his way, Cantore mentioned Jean Piaget and Werner Heisenberg as those who impressed him most deeply. He remembered all the support given by these scientists, particularly Heisenberg, who helped him in two ways, intellectually and spiritually: "I conducted the development-focused dialogue with scientists as part of my relations with the United Nations for the Conference mentioned above and during my trip to India, also mentioned." After all, Cantore followed Heisenberg's advice.

Later, in 1980, the Institute for Scientific Humanism was renamed the World Institute for Scientific Humanism (WISH). The institute never grew to the proportions Cantore had hoped for his envisioned influential body. According to a report in Fordham's *Ram* magazine, by 1979, the Institute boasted only seventeen members.[3] A conference paper from 1981 and a journal article from 1985 confirmed the existence of WISH. Still, they indicated that it was being reorganized, capturing Cantore's battle with finding ways to keep the momentum going at the Institute.[4] But the WISH never grew into the global magnet Cantore always imagined. In most respects, it was just that—a "wish," a grand idea that never became anything more than that.

By the late 1980s, the Institute was consigned to near oblivion. But all along the way, Cantore would never give up his dream of integrating science with humanism. For the next twenty years, from 1985 to 2006, Cantore continued writing his synthesis book, *Christ Wisdom and Science*, almost silently but quite diligently. As he approached the end of the book's writing, Cantore started thinking about the future and his return to Rome. "Now," he wrote, "I have no reason to stay in the United States, having completed the research project for which I came here. And so it seems important to me to devote my remaining years of good health to bring to fruition in the most fruitful way possible the mission that Christ has entrusted to me since way back in 1945."[5]

2. See Cantore, "My Humanistic-Scientific Apostolate."

3. See *Ram*, November 8, 1979.

4. Cantore, "Human Dignity, Science and Growth"; "Christic Origination of Science."

5. Cantore, "Un rapporto sul mio apostolato."

Cantore's was a journey of unwavering commitment to a project that sought to heal the divide between science and humanism. Though the World Institute for Scientific Humanism never came to have the vast influence he may have hoped for, Cantore's work did help open intellectual and spiritual dimensions in the lives of present-day scientists through the many institutes currently encouraging interdisciplinary dialogue in the sciences. To fuel their mission, we wish to remember Heisenberg's last words to Cantore, dated February 20, 1974: "I would hope that your new institute gives an opportunity to the younger generation to participate in fostering the use of science for building a more human world."

APPENDIX

The "Cantore Affair" at Fordham University

THE "CANTORE AFFAIR" is the complex and ultimately unsuccessful first attempt of Enrico Cantore to secure a solid academic standing and a specialized institute within Fordham University dedicated to scientific humanism. It unfolded between 1967 and 1968 and involved a series of communications and decisions by Fordham's academic and administrative authorities, resulting in the non-renewal of Cantore's contract.

In part, Cantore wanted to establish an Institute for Scientific Humanism at Fordham to heal the division between science and the humanities. Much of Cantore's work was interdisciplinary and focused on dialogue between scientists and humanists, making it new and strange. Still, it also had to be reconciled with the old structures of academic study found when he came to Fordham in 1966. For these reasons, many in Fordham's administration viewed Cantore, in the words of Lauer, as "an oddball" character. Among those most skeptical of Cantore were Quentin Lauer, Arthur Brown, and Patrick Heelan.

Although bold and creative, Cantore's vision was to prove contentious with the traditional structures and expectations of the university. At every turn, his search for an interdisciplinary institute that would heal the sciences and humanities met skepticism and concern from key figures in the institution. Problems such as his failure to teach, the practicality of his proposals, and his inability to provide enough resources to set up an institute caused increasing opposition.

Cantore also enlisted the support of leading scientists and theologians. Still, internal memoranda from those in charge at Fordham indicated a lack of confidence in his credentials to operate within the university's academic structure, so his appointment was not renewed. The relationship ended with Cantore being dismissed from Fordham early in 1968.

Whereas Father Enrico Cantore's letters read like a soul-baring, almost amorous memoir of his struggle to establish the Institute for Scientific Humanism at Fordham University, the university's internal documentation tells a different story.

This chapter is the result of extensive research work in the Archives of Fordham University, which tries to piece together this story from the viewpoint of Fordham's academic and administrative custodians.[1] The letters and memos of Fordham administrators reveal a different aspect of the story: institutional priorities, resource constraints, and concern for academic integrity played essential roles in the eventual decision to sever ties with Cantore.

We also reconstruct the events that led Werner Heisenberg to refuse an honorary doctorate from Fordham through internal communications and official responses. This perspective not only sheds light on the administrative and academic considerations shaping the outcome but also offers a fuller understanding of the Cantore Affair as a moment of tension between innovation and tradition within the university.

1. Cantore's Legacy and the Institute for Scientific Humanism

These papers preserve detailed accounts of the struggle of Enrico Cantore, whose dream was to build an academic institution. At the same time, this challenge is intellectual and moral, facing a new renaissance in which scientific and humanistic thought converge. Cantore had come to New York City in 1967 to take up an appointment as a faculty member in philosophy of science and coordinator of the Teilhard Research Institute at Fordham University.[2] His vision for the institute was ambitious:

---

1. Information about the original documents used to write this section can be provided by the author upon request.

2. In his "Faculty card record," Fr. Enrico Cantore is reported as assistant professor of philosophy of science (1967–1968), assistant professor of science/math L AC

he sought to make Fordham a hub for what he termed "scientific humanism," an intellectual movement to bridge the divide between the sciences and the humanities. In a December 1967 memorandum, sent only a few months after he arrived at Fordham, Cantore outlined his project, "How to Make Fordham a Creative Center of Scientific Humanism," addressing it to Arthur Brown, then dean of the Graduate School and later vice-president for academic affairs.[3]

In this memorandum, Cantore suggested that Fordham should begin by encouraging interdepartmental seminars and exchanges, harnessing the resources it already had in place academically to develop an interior dialogue that could, in time, reach outward. The second step in his plan would be to devise an institutional structure—a program center in the university to replace the Teilhard Research Institute, a headquarters for interdisciplinary dialogue with scientists. This was not just an academic initiative, but to Cantore, it was a necessary venture for any Christian university that wanted to respond to the intellectual crises of the contemporary world.

Following this first attempt, Cantore formulated a funding proposal entitled "Scientific Humanism at Fordham University: A Practical Program for Overcoming the Split Between the Two Cultures." This draft, probably in late 1967 or early 1968, set the record straight regarding what Fordham had already done in this direction, namely, the series of lectures given on Teilhard de Chardin's thought in 1963 by Maurits Huybens, then director of the *International Philosophical Quarterly*, and the workshops and conferences organized by the Teilhard Research Institute in the middle years of the decade. Cantore saw the ISH as a direct replacement for these efforts with monies already earmarked for the 1967–1968 academic year and scientific humanism taught in courses. The ISH is then presented as an event now operating, and Cantore describes himself in the following way:

> One faculty member with considerable experience in the interdisciplinary dialogue was freed from most of his teaching duties

---

(1970–1971), and adjunct assistant professor (1972–1973, 1974–1975, Fall 1976, 1977–Fall 1980). A note reports "Separated 1968 from regular faculty."

3. Dr. Arthur Brown resigned as vice president for academic affairs to accept the presidency of Marygrove College, a small liberal arts institution in Detroit. Dr. Brown was president of Adelphi College in Garden City, NY, before relocating to Fordham in 1967. He was the fourth president of Marygrove, and the second layman. See *Ram*, September 16, 1969, 5.

to organize the new structure. He has been consulting widely both with the various Fordham departments and with representatives of other institutions in the area of the Greater New York. To serve the cultural needs of this area it has been decided to establish the new Institute for Scientific Humanism. A budget of $8,000 has been appropriated for the year 1967/1968. Courses on Scientific Humanism have been instituted.

The activities he proposed for the ISH were, however, rather ambitious. Among them were the "Galileo Encounters," a series of lectures that included, for instance, the leading figure of the History of Science Department at Oxford, Dr. A. C. Crombie; and given their close intellectual relationship, there is little doubt that Cantore intended to invite Werner Heisenberg as well. The budget concluded with a detailed request for $29,000 over three years to support a small but dedicated staff to coordinate and organize these activities.

Over the years, and following his return to Fordham in 1969, Cantore continued advocating for ISH, updating the academic profile of the Institute and its goals. A document dated November 1973, revised in October 1977, presents a clear account of Cantore's academic background and outlines the beginning of ISH as official only in 1974. By this date, Cantore had written *Atomic Order* (1969, and paperback version in 1977) and *Scientific Man* (1977), and was scheduled to write another one: *Science and Dignity: The Challenge to Man*, which was not published.

Despite Cantore's efforts, the ISH needed assistance to become institutionally supported. A letter from May 5, 1978, addressed to Fordham's President James Finlay, shows that this was an ongoing issue. In this letter, Cantore presents the New York Forum for the Development of Man-Science, Technology, and Human Dignity, a project sponsored by the ISH and authorized by Joseph F. X. McCarthy, vice president for academic affairs. Finlay's response to this, which the letter mentions, alludes to his confusion about the university's function, clearly framing the marginal place the ISH had within the academic structure at Fordham.[4]

---

4. The letter introduces the New York Forum for the Development of Man-Science, Technology, and Human Dignity as an initiative promoted by the ISH, and approved by Joseph F. X. McCarthy—vice president for academic affairs. We read Finlay's bafflement, who circled the words "as a University undertaking" and added a reminder on the headline to ask McCarthy: "[Write] McC: What is this about? What is Fordham's involvement?" The attached document is a four-page summary presentation of the activity dated April 1978 containing a preliminary list of advisors (Alexander King, Aurelio Peccei, Klaus-Heinrich Standke, Philippe De Seynes, Rogers B. Finch, Stevan

Additional correspondence from November 1978 reveals more of Cantore's follow-ups to support ISH activities, especially a lecture series on "Global Development and Human Dignity." If anything, the university administrators' response could be characterized as lukewarm since Finlay ultimately declined the invitation to the lecture and the dinner afterward.

Thus, by May 1979, a memorandum by Thomas Massucci for the Fordham Faculty and Administrators again introduced ISH to the Fordham community, telling them about an institute incorporated in 1974. It speaks of Cantore's work to express the claims of scientific humanism in international fora, such as the UN Conference for Science and Technology for Development in Vienna that same year.[5] However, the tone of the document suggests that the ISH was still on the margins of academic life at Fordham and had not been able to gain a more solid position.

The last traces of the ISH still in existence, tucked away in the Fordham archives, are several items of promotional literature: leaflets, press releases, and draft articles. One of those is an article in Italian for a Jesuit magazine from September 18, 1979. It gives a retrospective on the ISH's journey, framing it as an "apostolic experience among scientists"—a strong reminder of Cantore's constancy to his spiritual vision of Scientific Humanism.

2. Correspondence and Events Leading to the Founding of the ISH

The letters to and from Cantore, scientists, and Fordham academic authorities dated between 1967 and 1968 offer an interesting picture of the initial excitement for Cantore's project, the reasons for the lack of support for the ISH, and the factors that contributed to the failure of the Institute to be set up as originally conceived. The archival papers, especially those in Leo McLaughlin's folder, tell a broader story of correspondence at the time, helping explain what was going on that led to the establishment and eventual collapse of Cantore's vision.

On December 14, 1967, Cantore sent a manuscript titled *Scientific Humanism and the Role of the University* to Leo McLaughlin, president of Fordham University. This manuscript appeared to outline the basic

---

Dedijer, Guy Gresford, Robert Muller, Aklilu Lemma, Eleonora Barbieri Masini).

5. The conference paper was later published as Cantore, "Human Dignity, Science and Growth."

idea of what Cantore hoped his institute would be, and the correspondence indicates that he and McLaughlin were actively exploring the potential project. By February 2, 1968, Cantore wrote to McLaughlin again, referring to a conversation two weeks earlier, during which he was encouraged by McLaughlin to seek more comprehensive support from Fordham's scientific faculty. He included a copy of a memorandum he had previously sent to Arthur Brown, dean of the Graduate School, outlining his proposal for the ISH. Cantore requested that this be discussed at an interdepartmental meeting on February 29, 1968.

Encouraged by Cantore's initiative, numerous scientists and intellectuals wrote to McLaughlin expressing their support. Francisco J. Ayala of Rockefeller University and Alberto Dou, SJ, of Madrid were among the enthusiastic supporters. Christopher Mooney, SJ, however, expressed caution, questioning whether such a project aligned with Fordham's academic mission. McLaughlin's responses to these letters on February 8, 1968, suggested a reluctance to fully commit to the project.

In the midst of these exchanges, Angelo Serra, SJ, from the Institute of Human Genetics at the Catholic University of the Sacred Heart in Rome wrote a very enthusiastic letter. Still, it went unanswered—possibly indicating a shift in the university's stance or disagreement with Cantore's proposal. Peter Berger from the New School for Social Research also expressed support, but like several others, his letter received only a brief acknowledgment from McLaughlin.

A key letter came from Michael Yanase, SJ, of the Institute for Advanced Studies at Princeton on February 13, 1968. He emphasized that Cantore's program aligned with the spirit of Vatican II and urged caution in handling the project. However, in a personal note, Yanase questioned whether Cantore was the right person to lead the initiative, hinting at concerns within Fordham's administration.

The final letter in this early exchange came from Thomas B. Settle of Brooklyn Polytechnic Institute on February 16, 1968. McLaughlin's reply on February 20 referred to "complications" that had arisen, signaling internal discussions at Fordham were steering away from supporting Cantore's initiative.

The following document was a draft letter McLaughlin planned to send to the scientists who supported the ISH, expressing regret that the proposal could not move forward due to a lack of staff. He also noted that while the administration was initially interested, practical limitations made it impossible to proceed. McLaughlin added that Cantore would be leaving Fordham at the end of the summer to return to the Gregorian University in Rome, effectively closing the door on the ISH at Fordham.

On February 27, 1968, McLaughlin sent this final decision to all the recipients, including Francisco Ayala. The tone of the correspondence shifted from cautious optimism to finality, signaling the end of a promising but ultimately unworkable proposal.

Reactions to McLaughlin's February 27 letter revealed disappointment among the scientists. On March 18, 1968, Peter Berger expressed frustration, noting that Cantore's proposal had merit but had two major flaws. First, Berger felt the proposal took the divide between the sciences and the humanities too seriously, potentially widening rather than bridging the gap. Second, he criticized Cantore for excluding theological participation from the institute, a decision Berger found troubling given the interdisciplinary nature of the project. As a non-Catholic layman, Berger emphasized the need for genuine interdisciplinary dialogue without rigid boundaries between fields.

More pointed was the response from Angelo Serra, SJ, on April 13, 1968. Serra lamented what he saw as a "lack of sincerity" throughout the affair, indicating that trust and transparency were ultimately lacking in evaluating Cantore's proposal: "At any rate, God has his own ways, and leads men willingly unwillingly along them," he wrote. Though brief, Serra's response hinted at deeper tensions that may have foreshadowed the eventual collapse of the ISH initiative.

This correspondence paints a complex picture of Cantore's struggles to establish the ISH at Fordham. While it initially showed promise, with support from both the faculty and external scientists, the initiative ultimately faltered due to internal skepticism, logistical challenges, and perhaps a disconnect between Cantore's vision and the university's goals. The outcome led to Cantore's departure from Fordham, ending his brief tenure at the institution.

3. The Failure to Award Werner Heisenberg an Honorary Doctorate

This section will attempt to reconstruct the events surrounding the failed attempt to award Werner Heisenberg an honorary doctorate from Fordham University in 1968.[6] Cantore suggested that the degree

---

6. See Letters 25–27.

be awarded in "Humane Letters" rather than science, as he believed it would better reflect Heisenberg's work in bridging the gap between science and humanism.

On February 6, 1968, Enrico Cantore wrote a letter to Leo McLaughlin addressing an important issue regarding Werner Heisenberg's honorary degree. Cantore, who had been instrumental in initiating this honor,[7] requested that Heisenberg be awarded the degree of Doctor Honoris Causa in "Humane Letters" rather than an honorary degree in science. Cantore explained that the initial offer had mistakenly proposed the latter due to a secretarial error. He argued that only the title of Humane Letters could appropriately recognize and emphasize Heisenberg's significant contributions to philosophical thought, which were crucial for bridging the perceived divide between the sciences and the humanities. To support his request, Cantore enclosed a letter from Heisenberg, in which the renowned physicist expressed his gratitude for the proposed honorary doctorate.[8]

McLaughlin responded promptly on February 8, 1968, the same day he had been corresponding with several scientists regarding Cantore's larger project for the Institute for Scientific Humanism. In his reply, McLaughlin acknowledged Cantore's request and saw no issue with changing the title of the honorary degree to Doctor Honoris Causa in Humane Letters. He also inquired whether he could inform Heisenberg of this modification, indicating a willingness to proceed with Cantore's recommendation.[9]

However, a further complication arose on February 16, 1968, when Timothy Healy, the executive vice president of Fordham, wrote to Cantore. In his letter, Healy referred to a conversation with Cantore the previous day and provided a crucial clarification regarding the conditions of the honorary doctorate offer. Healy reminded Cantore that the offer to confer an honorary doctorate upon Heisenberg was contingent upon the latter's presence at the official University Commencement in June of that year. He further emphasized that if Heisenberg were unable to attend, the offer would simply lapse, as the university only awarded honorary degrees during its commencement ceremony. Healy also noted that while

---

7. See Letter 14.
8. See Letters 16 and 18.
9. See Letter 20.

Heisenberg might be considered again in the future, no promises or assurances could be made about re-nomination for the following year.

Healy's message, though procedurally sound, carried a tone of finality that may have reflected deeper concerns within the university about Cantore's initiatives. Interestingly, Healy's letter did not mention the forthcoming decision regarding Cantore's position at Fordham, which would be communicated just a few days later, on February 19, 1968. This leaves a strong impression that, while Cantore was focused on securing academic recognition for Heisenberg, the administration was simultaneously making decisions about his future at Fordham.

In retrospect, the discussions around Heisenberg's honorary doctorate may have been just the tip of the iceberg, pointing to deeper tensions that defined Cantore's relationship with the university. The procedural concerns raised in Healy's letter suggest a broader institutional caution that influenced not only the awarding of honors but also the direction of Cantore's career at Fordham. Whether due to Cantore's dismissal or other reasons, Heisenberg ultimately declined the offer of the honorary doctorate on April 2, 1968.[10]

4. The Dismissal of Enrico Cantore from Fordham University

The most decisive and dramatic moment in the Cantore Affair occurred in 1968: his dismissal from Fordham. This section outlines the events leading to that decision, the roles played by key figures such as Timothy Healy, Leo McLaughlin, and Arthur Brown, and the broader institutional dynamics. Internal documents from Leo McLaughlin's files provide a clear account of the circumstances surrounding Cantore's departure.

The first significant communication about Cantore's position at Fordham was a letter dated January 31, 1968, from Timothy Healy, the executive vice president of Fordham, to Leo McLaughlin. Healy informed McLaughlin that Francis J. McCool, a delegate of the Jesuit Curia in Rome, had inquired about the possibility of assigning Cantore to Fordham permanently. This inquiry was prompted by two enthusiastic letters from Cantore, one in October 1967 and another in January 1968, in which he offered a highly optimistic assessment of his work at the university. McCool sought feedback from Quentin Lauer (chairman of the Philosophy Department), Arthur Brown (vice president for

10. See Letter 30.

academic affairs), and Patrick Heelan (professor) regarding Cantore's performance and the need for his continued service.

It's important to note that Cantore had only recently arrived at Fordham from Rome. His early push for a permanent appointment likely put the university administration in a difficult position, as they were being asked to make a long-term commitment without sufficient time to fully evaluate the impact of his work. This premature pressure contributed to concerns that would surface later.

In his January 31 letter, Healy reported that both Lauer and Brown expressed negative views about retaining Cantore. The primary issue was Cantore's apparent desire to reduce his teaching responsibilities. Although assigned two courses, Cantore managed to reduce his load to one by privately negotiating with the dean of the Business School to withdraw from teaching without informing Lauer until the fact was accomplished. Additionally, there was a belief that the Teilhard Research Institute, which Cantore was coordinating, could function effectively without his involvement. Based on these factors, Healy concluded that Cantore's services were not essential to Fordham at that time or in the foreseeable future.

Following these verbal assessments, McLaughlin requested written confirmations from Brown and Lauer to finalize the decision not to retain Cantore. These letters were necessary for McLaughlin to proceed with any formal actions regarding Cantore's position at the university. Based on side notes in pencil on the attached document, there were prior verbal agreements, especially with Lauer, that were later formalized in writing.

On January 26, 1968, McCool contacted Patrick Heelan, copying Healy, Brown, and Lauer. Heelan's response, dated January 31, 1968, was frank and supportive of Cantore's work but not without reservations. Heelan emphasized the importance of Cantore's interdisciplinary efforts, which at that time were considered a "para-academic" field in the American academy. Heelan provided a detailed evaluation of Cantore's role within the faculty, noting that Cantore had reluctantly accepted teaching responsibilities due to ambiguity in his original invitation to Fordham. Heelan also noted that Cantore did not align with contemporary trends in the philosophy of science and presented himself as an "interdisciplinary catalyst for exchanges between scientists and humanists"—a role that had no clear place in academic structures of the time. Heelan doubted whether the university could fund Cantore's work and ultimately agreed with Lauer that offering Cantore a permanent position at Fordham would be premature.

While Heelan's assessment recognized the value of Cantore's work, it highlighted the disconnect between Cantore's ambitions and Fordham's institutional capabilities. His letter, also sent to McLaughlin, significantly shaped the final decision.

The decisive moment in the Cantore affair came with a letter from Emil Moriconi, chairman of the Chemistry Department, to Healy on February 5, 1968. Moriconi expressed frustration with how Cantore's presence at Fordham had been managed. He recounted how, in the interest of fostering interdepartmental cooperation, he had spent considerable time discussing matters with Cantore and had even encouraged two faculty members, Hennessy and Franck, to attend Cantore's seminars, assuming these requests were known to the Philosophy Department.

Moriconi was quite alarmed to learn from internal sources that Cantore was trying to establish the headquarters of the Institute for Scientific Humanism in the Chemistry Department building and had even proposed himself as a department member despite lacking the necessary qualifications. Moriconi strongly opposed this and threatened to resign if the request was approved without proper consultation. He emphasized that the Chemistry Department did not have the space or resources for Cantore's proposed activities and had no intention of supporting the request.

On February 19, 1968, Arthur Brown informed Cantore that his contract with Fordham for 1968–1969 would not be renewed, effectively ending his brief tenure at the university.

Glossary of Names

ANSHEN, RUTH NANDA (1900–2003). American philosopher. Author and editor at Haper & Row, New York.

ARRUPE GONDRA, PEDRO (1907–1991). Spanish Basque Jesuit who served as the 28th superior general of the Society of Jesus from 1965 to 1983.

AUTRUM, HANSJOCHEM (1907–2003). Chairman, Department of Zoology, University of Munich.

AYALA, FRANCISCO J. (1934–2023). Spanish-American evolutionary biologist and philosopher. Professor at the University of California, Irvine, and University of California, Davis.

BAŁCZEWSKI, KAZIMIERZ. Polish Jesuit, philosopher, and physicist.

BARBIERI MASINI, ELEONORA (1928–2022). Italian sociologist, one of the founders of contemporary futurology. In 1975 she was appointed secretary of the World Futures Studies Federation and joined the Club of Rome. The following year she was appointed to teach human and social foresight at the Pontifical Gregorian University, to which she was added chair of human ecology in 1991.

BERGER, PETER (1929–2017). Austrian-born American sociologist and Protestant theologian.

BISCHOF, NORBERT (born 1930). German behavioral psychologist and system theorist. He worked with Konrad Lorentz at MPI for Behavioral Psychology in Seewiesen (1966–1975).

GLOSSARY OF NAMES

BOHR, NIELS (1885–1962). Danish physicist, philosopher, and a promoter of scientific research. Nobel in Physics (1922).

BORN, MAX (1882–1970). German-British physicist and mathematician. Contributed to the development of quantum mechanics, solid-state physics, and optics. Nobel in Physics (1954).

BOWEN, CARROLL. MIT Press's first director.

BRENDEL, BETTINA (1922–2009). German-born Los Angeles painter and educator.

BRONK, DETLEV WULF (1897–1975). American scientist, educator, and administrator. President of Johns Hopkins University (1949–1953) and president of Rockefeller University (1953–1968).

BROWN, ARTHUR (1918–2006). Vice president for academic affairs at Fordham University. Later became president of Mary Grove College in Detroit. Dean of Baruch College, CUNY. Dean of arts and sciences at the University of Miami in Miami.

BÜCHEL, WOLFGANG (1920–1990). German Jesuit and philosopher who was particularly interested in the philosophy of physics.

BURDICK, BERNARD JOSEPH (born 1941). PhD graduate student in physics, Case Western Reserve University.

CHAKRAVARTY, AMIYA (1901–1986). Professor of philosophy at the State University of New York, New Paltz, NY.

CHULLIKAL, ANTHONY. Philosopher and economist. International Commission for Justice and Peace, Rome.

COURANT, RICHARD (1888–1972). German American mathematician.

COX, HARVEY GALLAGHER JR. (born 1929). American theologian, Hollis Professor of Divinity at Harvard Divinity School, until his retirement in October 2009. Author of *The Secular City* (1965)

CROMBIE, ALISTAIR CAMERON (1915–1996). Australian historian of science who began his career as a zoologist. Head of the History of Science Department at the University of Oxford.

DEDIJER, STEVAN (1911–2004). Yugoslav physicist and pioneer of business intelligence.

DOBZHANSKY, THEODOSIUS GRIGORIEVICH (1900–1975). Ukrainan-American geneticist and evolutionary biologist. Professor at Rockefeller University (1962–1971).

DOLCH, HEIMO (1912–1984). Professor of philosophy at the University of Bonn.

DOU, ALBERTO (1915–2009). Spanish Jesuit and mathematician.

DÜRR, HANS-PETER (1929–2014). German physicist. Professor of physics at the Ludwig Maximilian University, Munich, Germany. Vice executive director at the Max Planck Institute for Physics 1972–1977, 1981–1986, 1993–1995.

FINCH, ROGERS B. (1920–2006). Vice president of the World Federation of Engineering Organizations. Officer and board member of the American Society of Association Executives.

FOLEY, EUGENE P. (1928–2015). Investment businessman with an education in philosophy and law. Treasurer and financial chairman of ISH.

GIESE, ANNEMARIE. Werner Heisenberg's secretary.

GONSETH, FERDINAND (1890–1975). Swiss mathematician and philosopher.

GRAHAM, LOREN R. (born 1933). American historian of science, particularly science in Russia. Professor emeritus of the history of science, Harvard University.

GRESFORD, GUY B. Department for Economic and Social Affairs, United Nations University.

HABACHI, RENÉ-MARC (1915–2003). Director, UNESCO Division of Philosophy, Paris.

HAYES F. Translator of *Philosophical Problems of Quantum Physics* for Faber and Faber.

HEALY, TIMOTHY (1923–1992). Jesuit priest and academic who held several prominent positions, including executive vice president at Fordham University. He later served as vice chancellor for academic affairs at the City University of New York (CUNY) and as president of Georgetown University. Healy was also president of the New York Public Library. Born on April 25, 1923, he passed away on December 30, 1992.

HEELAN, PATRICK A. (1926–2015). Irish Jesuit theoretical physicist and philosopher. Professor of philosophy at Georgetown until 2013.

HEISENBERG-SCHUMACHER, ELISABETH (1914–1998). Werner Heisenberg's wife.

HODGSON, PETER EDWARD (1928–2008). Senior research fellow and lecturer of physics, Corpus Christi College, University of Oxford.

HOPP, FRIEDRICH ARNOLD "FRITZ" (1909–1987). German theoretical physicist who contributed to nuclear physics. Professor of theoretical physics at the University of Munich.

HUYBENS, JAN-MAURITS (1920–2005). Belgian Jesuit. Taught of philosophy at the Jesuit faculty in Leuven and Heverlee (1957–1969). European editor of the *International Philosophical Quarterly* (1960–1968).

HYNEK J. ALLEN (1910–1983). Chairman, Department of Astronomy, Northwestern University, Evanston, Illinois

JAMMER, MAX (1915–2010). Israeli physicist and philosopher of physics.

JEUKEN, MARIUS (1916–1983). Dean, Department of Biology and director, Institute for Theoretical Biology, University of Leiden, Holland.

KERBER, WALTER (1926–2006). German Jesuit, social ethicist, and Catholic theologian.

KING, ALEXANDER (1909–2007). British chemist and pioneer of the sustainable development movement who co-founded the Club of Rome in 1968. Director-general for scientific affairs at the Organisation for Economic Co-operation and Development (OECD).

KOMAR, ARTHUR B. (1931–2011) Professor of physics and dean, Belfer Graduate School of Science, Yeshiva University, New York City.

LAUER, QUENTIN (1917–1997). Jesuit philosopher and scholar who served as the chairman of the Philosophy Department at Fordham University.

LEMMA, AKLILU (1935–1997). Ethiopian pathologist. He served the UN in many capacities as a scientist, became the deputy director of UNICEF's International Child Development Centre.

MAGNUS, HANS HEINRICH WILHELM (1907–1990). German-American mathematician. Professor of mathematics at the Polytechnic Institute of New York, Brooklyn, NY.

MARGENAU, HENRY (1901–1997). German-American physicist and philosopher of science.

MATHIAS, JAMES F. Administrative vice president and secretary, Guggenheim Memorial Foundation

MCCARTHY, JOSEPH F. X. (1922–2012). Vice president of academic affairs of Fordham University from 1976 to 1984, was also a professor at the Graduate School of Education.

MCCOOL, FRANCIS J. (1913–1996). Scholar at the Pontifical Biblical Institute in Rome. Delegate of the Jesuit Interprovincial Curia in Rome.

MCLAUGHLIN, LEO (1912–1996). President of Fordham University during 1965–1968.

MITTELSTAEDT, HORST (1923–2016). German biologist and cybernetician. Until 1999 he worked at the Max Planck Institute for Behavioral Physiology at Seewiesen, Bavaria.

MONOD, JACQUES LUCIEN (1910–1976). French biochemist, Nobel in Physiology or Medicine in 1965.

MOONEY, CHRISTOPHER (1925–1993). Jesuit priest, lawyer, theologian. Professor and chair of the theology department at Fordham University from 1964 to 1969.

MORICONI, EMIL J. Chairman and professor of the Chemistry Department at Fordham University.

MULLER, ROBERT (1923–2010). Prominent theosophist and international globalist who served with the United Nations for forty years, ultimately reaching the rank of assistant secretary-general.

PAIS, ABRAHAM (1918–2000). Dutch-American physicist and science historian. Professor of physics at Rockefeller University until his retirement.

PAULI, WOLFGANG ERNST (1900–1958). Austrian theoretical physicist. Nobel in Physics (1945).

PECCEI, AURELIO (1908–1984). Economist and founder of the Club of Rome, Rome

PIAGET, JEAN WILLIAM FRITZ (1896–1980). Swiss psychologist. Piaget's theory of cognitive development and epistemological view are together called "genetic epistemology."

POMERANS, ARNOLD JULIUS (1920–2005). German-born British translator. Translator of *Physics and Beyond* for Haper and Row.

RABI, ISIDOR ISAAC (1898–1988). American physicist, Nobel in Physics (1944). Professor at Columbia University of New York.

REISS, PAUL JACOB (born 1930). Administrator of Fordham University. Professor of sociology, both executive vice president for Lincoln Center and dean of the Liberal Arts College at Lincoln Center.

RONCHI, VASCO (1897–1988). Director, National Institute of Optics, Florence. Former president, International Union of History and Philosophy of Science.

SALVIUCCI, PIERO (1899–1982). Former chancellor of the Pontifical Academy of the Sciences.

SCHEIBE ERHARD (1927–2010). German philosopher of science. Professor of philosophy at the University of Göttingen

SEEGER, RAYMOND JOHN (1906–1992). American physicist. Deputy assistant director and senior staff research associate at the National Science Foundation (1952–1970). Professor at the American University from 1954 to 1972.

SEITZ, FREDERICK (1911–2008). American physicist. President of Rockefeller University (1968–1978).

SERRA, ANGELO (1919–2012). Italian Jesuit and world-renowned geneticist expert on trisomy-21 (Down syndrome). Professor of genetics and founder of the Institute and chair of the Human Genetics Department at the Catholic University of Rome School of Medicine. Member of the Pontifical Academy for Life.

SETTLE, THOMAS B. (1930–2020). Scholar of early modern history, specialist in Galileo Galilei.

SEYNES, PHILIPPE DE (1910–2003). French economist. Head of the UN Economic and Social Department for twenty years.

SHEA, GEORGE WILLIAM (born 1934). Associate professor of classics and Dean, Liberal Arts College, Fordham University at Lincoln Center, New York City

SHEAHAN, MICHAEL U. (1923–2012). Board of Lay Trustees, Fordham University.

SITTLER, JOSEPH ANDREW (1904–1987). American Lutheran minister and theologian who taught at Maywood Seminary (Divinity School of the University of Chicago) and the Lutheran School of Theology at Chicago.

SKINNER, BURRHUS FREDERIC (1904–1990). American psychologist, behaviorist, author, inventor, and social philosopher. Professor of psychology at Harvard University (1958–1974).

SMITH, CYRIL STANLEY (1903–1992). British metallurgist and historian of science. Professor in the Departments of Humanities and Metallurgy at MIT.

SMITH, JOHN EDWIN (1921–2009). American philosopher and Clark Professor of Philosophy at Yale University. He served as president of the American Philosophical Society.

STANDKE, KLAUS-HEINRICH (born 1935). German economist and international science policy expert in international governmental organizations (OECD & EIRMA, Paris; United Nations, New York; UNESCO, Paris). He was entrusted with the preparation of a UN world conference, the United Nations Conference for Science and Technology for Development (UNCSTD), Vienna, August 20–31, 1979, and the World Science Forum "Science, Technology and Society. Needs, Challenges and Limitations," Vienna, August 13–17, 1979.

SÜSSMANN, G. (1928–2017). Professor of theoretical physics at the University of Munich.

TRINKLEIN, FREDERICK ERNST (1924–2002). Science educator and writer with special training in physics and astronomy. Author of *The God of Science*. Deputy Director of ISH.

TRINKLEIN, MARGARET (born 1926). Teacher on the primary and secondary levels. Wife of Fredrick Trinklein. Secretary of ISH.

UHLENBECK, GEORGE EUGENE (1900–1988). Dutch-American theoretical physicist. Professor of physics (1960–1988) at Rockefeller University.

WARNOW-BLEWETT (NÉE NELSON), JOAN (1931–2006). Librarian, American Institute of Physics.

WATANABE, SATOSI (1910–1993). Japanese theoretical physicist. Researcher at the IBM Watson Laboratory, taught at Yale University and University of Hawaii.

WEISSKOPF, VICTOR (1908–2002). Austrian-born American theoretical physicist. Professor of physics at MIT.

WEIZSÄCHER, CARL FRIEDERICH (VON) (1912–2007). German physicist, astrophysicist, and philosopher. Professor of philosophy at the University of Hamburg.

WHEELER, JOHN ARCHIBALD (1911–2008). American theoretical physicist. Professor of physics at Princeton University (1938–1976).

YANASE, MICHAEL MUTSUO (1928–1991). Jesuit-physicist, member of the School of Natural Sciences at the Institute for Advanced Studies, Princeton.

YOURGRAU, WOLFGANG H. J. (1908–1979). Professor of history and philosophy of science at the University of Denver, Colorado (1963–1978).

Bibliography

Archival Sources

Archives of the Max Plank Society (Archiv der Max-Planck-Gesellschaft), Berlin

Folders consulted: Heisenberg—W. Büchel (III. Abt., Rep. 93, Nr. 44); Heisenberg—W. Wickler (III. Abt., Rep. 93, Nr. 158); Heisenberg—P. A. Heelan (III. Abt., Rep. 93, Nr. 204); Heisenberg's single manuscripts (Manuskripte eigene) (III. Abt., Rep. 93, Nr. 1014); Annemarie Giese (III. Abt., Rep. 93, Nr. 1661); Heisenberg—R. N. Anshen (III. Abt., Rep. 93, Nr. 1711); Heisenberg—Cantore (III. Abt., Rep. 93, Nr. 1724); Heisenberg—Fordham University (III. Abt., Rep. 93, Nr. 1872); Heisenberg—A. Giese (III. Abt., Rep. 93, Nr. 1971).

Letter 1. Signature: III/93/1724/493–96, Mappe 4. Brief von Enrico Cantore von Pontificia Università Gregoriana an Werner Heisenberg. Language: German.

Letter 2. Signature: III/93/1724/491–92, Mappe 4. Brief von Pietro Salviucci von Pontificia Accademia delle Scienze (Vatikanstadt) an Werner Heisenberg. Language: German.

Letter 3. Signature: III/93/1724/489–90, Mappe 4. Brief von Annemarie Giese von Max-Planck-Institut für Physik. Direktion. Sekretariat an Enrico Cantore an Pontificia Università gregoriana. Language: German.

Letter 4. Signature: III/93/1971/171–72, Brief von Enrico Cantore an Annemarie Giese. Language: German.

Letter 5. Signature: III/93/1724/483–88, Mappe 4. Brief von Enrico Cantore von Fordham University an Werner Heisenberg. Language: German.

Letter 6. Signature: III/93/1724/457–70, Mappe 4. Brief von Enrico Cantore von Fordham University an Annemarie Giese an Max-Planck-Institut für Physik und Astrophysik. Language: German.

Letter 7. Signature: III/93/1724/479–82, Mappe 4. Brief von Werner Heisenberg an Enrico Cantore an Fordham University. Language: German.

Letter 8. Signature: III/93/1724/473–78, Mappe 4. Brief von Enrico Cantore von Fordham University an Werner Heisenberg an Max-Planck-Institut für Physik und Astrophysik. Language: German.

Letter 9. Signature: III/93/1724/471–72, Mappe 4. Brief von Werner Heisenberg an Enrico Cantore an Fordham University. Language: German.

Letter 10. Signature: III/93/1724/455–56, Mappe 4. Brief von Enrico Cantore von Fordham University an Werner Heisenberg an Max-Planck-Institut für Physik und Astrophysik. Language: German (attachment in English).

Letter 11. Signature: III/93/1724/453–54, Mappe 4. Brief von Werner Heisenberg an Enrico Cantore an Fordham University. Language: German.

Letter 12. Signature: III/93/1724/451–52, Mappe 4. Brief von Enrico Cantore von Fordham University an Werner Heisenberg an Max-Planck-Institut für Physik und Astrophysik. Language: German.

Letter 13. Signature: III/93/204/201–2, Brief von Werner Heisenberg an Patrick A. Heelan an Fordham University. Language: English.

Letter 14. Signature: III/93/1724/447–50, Mappe 4. Brief von Enrico Cantore von Fordham University an Werner Heisenberg. Language: German.

Letter 15. Signature: III/93/1724/445–46, Mappe 4. Brief von Enrico Cantore von Fordham University an Werner Heisenberg an Max-Planck-Institut für Physik und Astrophysik. Language: German.

Letter 16. Signature: III/93/1724/443–44, Mappe 4. Brief von Werner Heisenberg an Enrico Cantore an Fordham University. Language: German.

Letter 17. Signature: III/93/1872/397–98, Mappe 2. Brief von Leo McLaughlin von Fordham University an Werner Heisenberg an Max-Planck-Institut für Physik und Astrophysik. Language: English.

Letter 18. Signature: III/93/1724/441–42, Mappe 4. Brief von Enrico Cantore von Fordham University an Werner Heisenberg an Max-Planck-Institut für Physik und Astrophysik. Language: German.

Letter 19. Signature: III/93/1724/439–40, Mappe 4. Brief von Werner Heisenberg an Enrico Cantore an Fordham University. Language: German.

Letter 20. Signature: III/93/1872/395–96, Mappe 2. Brief von Fordham University an Werner Heisenberg an Max-Planck-Institut für Physik und Astrophysik. Language: English.

Letter 21. Signature: III/93/1724/421–34, Mappe 4. Brief von Enrico Cantore von Fordham University an Werner Heisenberg. Language: German (attachment in English).

Letter 22. Signature: III/93/1724/435–38, Mappe 4. Brief von Enrico Cantore von Fordham University an Werner Heisenberg an Max-Planck-Institut für Physik und Astrophysik. Language: English.

Letter 23. Signature: III/93/1872/393–94, Mappe 2. Brief von Werner Heisenberg an Leo McLaughlin an Fordham University. Language: English.

Letter 24. Signature: III/93/1872/391–92, Mappe 2. Brief von Leo McLaughlin von Fordham University an Werner Heisenberg an Max-Planck-Institut für Physik und Astrophysik. Language: English.

Letter 25. Signature: III/93/1724/417–20, Mappe 3. Brief von Wolfgang Büchel von Berchmanskolleg (Pullach i. Isartal) an Werner Heisenberg. Language: German (attachment in German).

Letter 26. Signature: III/93/1724/407–16, Mappe 3. Brief von Enrico Cantore von Fordham University an Werner Heisenberg an Max-Planck-Institut für Physik und Astrophysik. Language: English (attachment not included).

Letter 27. Signature: III/93/1724/403-4, Mappe 3. Brief von Werner Heisenberg an Enrico Cantore an Fordham University. Language: German.
Letter 28. Signature: III/93/1724/405-6, Mappe 3. Brief von Werner Heisenberg an Detlev Wulf Bronk an Rockefeller University. Language: English.
Letter 29. Signature: III/93/44/183-84, Brief von Werner Heisenberg an Wolfgang Büchel an Berchmanskolleg (Pullach i. Isartal), 02.04.1968. Language: German.
Letter 30. Signature: III/93/1872/389-90, Mappe 2. Brief von Werner Heisenberg an Fordham University, 02.04.1968. Language: English.
Letter 31. Signature: III/93/1724/401-2, Mappe 3. Brief von Enrico Cantore von Fordham University an Werner Heisenberg. Language: German.
Letter 32. Signature: III/93/1724/399-400, Mappe 3. Brief von Enrico Cantore von Fordham University an Werner Heisenberg an Max-Planck-Institut für Physik und Astrophysik. Language: English.
Letter 33. Signature: III/93/1724/397-98, Mappe 3. Brief von Werner Heisenberg an Enrico Cantore. Language: German.
Letter 34. Signature: III/93/1724/395-96, Mappe 3. Brief von Enrico Cantore von Heythrop College an Werner Heisenberg. Language: English.
Letter 35. Signature: III/93/1724/393-94, Mappe 3. Brief von Annemarie Giese von Max-Planck-Institut für Physik. Direktion. Sekretariat an Enrico Cantore an Heythrop College. Language: German.
Letter 36. Signature: III/93/1724/391-92, Mappe 3. Brief von Enrico Cantore von Heythrop College an Werner Heisenberg. Language: English.
Letter 37. Signature: III/93/1724/309-88, Mappe 3. "Scientific Humanism and the Role of the University" (Manuskripttitel). Language: English ["Principles for Scientific Humanism"] (two papers in attachment not included).
Letter 38. Signature: III/93/1724/389-90, Mappe 3. Brief von Werner Heisenberg an Enrico Cantore an Heythrop College. Language: German.
Letter 39. Signature: III/93/1724/301-2, Mappe 2. Brief von Enrico Cantore von Heythrop College an Werner Heisenberg. Language: English.
Letter 40. Signature: III/93/1711/333-34, Mappe 1. Brief von Werner Heisenberg an Ruth Nanda Anshen. Language: German.
Letter 41. Signature: III/93/1724/303-8, Mappe 2. Brief von Enrico Cantore von Heythrop College an Werner Heisenberg. Language: English.
Letter 42. Signature: III/93/1971/163-64, Brief von Annemarie Giese von Max-Planck-Institut für Physik. Direktion. Sekretariat an Norbert Bischof an Max-Planck-Institut für Verhaltensphysiologie. Language: German.
Letter 43. Signature: III/93/1711, Mappe 1. Brief von Ruth Nanda Anshen an Werner Heisenberg. Language: English.
Letter 44. Signature: III/93/1724/299-300, Mappe 2. Brief von Enrico Cantore von Heythrop College an Werner Heisenberg. Language: English.
Letter 45. Signature: III/93/1724/297-98, Mappe 2. Brief von Werner Heisenberg an Enrico Cantore. Language: German.
Letter 46. Signature: III/93/1711, Mappe 1. Brief von Ruth Nanda Anshen an Werner Heisenberg. Language: English.
Letter 47. Signature: III/93/1711/325-28, Mappe 1. Brief von Werner Heisenberg an Ruth Nanda Anshen. Language: German.
Letter 48. Signature: III/93/1724/295-96, Mappe 2. Brief von Enrico Cantore an Werner Heisenberg. Language: English.

Letter 49. Signature: III/93/1724/293–94, Mappe 2. Brief von Werner Heisenberg an Enrico Cantore. Language: German.

Letter 50. Signature: III/93/1711, Mappe 1. Brief von Ruth Nanda Anshen an Werner Heisenberg. Language: English.

Letter 51. Signature: III/93/1724/289–92, Mappe 2. Brief von Enrico Cantore an Werner Heisenberg. Language: English.

Letter 52. Signature: III/93/1711, Mappe 1. Brief von Werner Heisenberg an Ruth Nanda Anshen. Language: German.

Letter 53. Signature: III/93/1724/287–88, Mappe 2. Brief von Werner Heisenberg an Frederick Seitz an Rockefeller University. Language: English.

Letter 54. Signature: III/93/1711, Mappe 1. Brief von Ruth Nanda Anshen an Werner Heisenberg. Language: English.

Letter 55. Signature: III/93/1724/279–80, Mappe 2. Brief von Enrico Cantore an Werner Heisenberg. Language: English.

Letter 56. Signature: III/93/1724/281–82, Mappe 2. Brief von Frederick Seitz von Rockefeller University an Werner Heisenberg an Max-Planck-Institut für Physik und Astrophysik. Language: English.

Letter 57. Signature: III/93/1724/277–78, Mappe 2. Brief von Enrico Cantore an Werner Heisenberg an Max-Planck-Institut für Physik und Astrophysik. Language: English.

Letter 58. Signature: III/93/1724/285–86, Mappe 2. Brief von Joan Warnow-Blewett von American Institute of Physics an Werner Heisenberg an Max-Planck-Institut für Physik und Astrophysik. Language: English.

Letter 59. Signature: III/93/1724/283–84, Mappe 2. Brief von Werner Heisenberg an Joan Warnow-Blewett an American Institute of Physics. Language: English.

Letter 60. Signature: III/93/1711, Mappe 1. Telegram von Ruth Nanda Anshen an Werner Heisenberg. Language: English.

Letter 61. Signature: III/93/1711, Mappe 1. Brief von Werner Heisenberg an Ruth Nanda Anshen. Language: German.

Letter 62. Signature: III/93/1711, Mappe 1. Brief von Ruth Nanda Anshen an Werner Heisenberg. Language: English.

Letter 63. Signature: III/93/1711, Mappe 1. Brief von Werner Heisenberg an Ruth Nanda Anshen. Language: German.

Letter 64. Signature: III/93/1711, Mappe 1. Brief von Werner Heisenberg an Ruth Nanda Anshen. Language: German (attachment not included).

Letter 65. Signature: III/93/1724/275–76, Mappe 2. Brief von Enrico Cantore an Werner Heisenberg an Max-Planck-Institut für Physik und Astrophysik. Language: English.

Letter 66. Signature: III/93/1724/273–74, Mappe 2. Brief von Werner Heisenberg an Enrico Cantore. Language: German.

Letter 67. Signature: III/93/1724/259–72, Mappe 2. Brief von Enrico Cantore an Werner Heisenberg. Language: English (attachment in English).

Letter 68. Signature: III/93/1971/159–60, Brief von Enrico Cantore an Annemarie Giese. Language: German.

Letter 69. Signature: III/93/1724/255–58, Mappe 2. Brief von Werner Heisenberg an Enrico Cantore. Language: English (attachment in English).

Letter 70. Signature: III/93/1724/253-54, Mappe 2. Brief von Enrico Cantore an Werner Heisenberg an Max-Planck-Institut für Physik und Astrophysik. Language: English.

Letter 71. Signature: III/93/1724/249-52, Mappe 2. Brief von Enrico Cantore an Werner Heisenberg. Language: English (attachment in English).

Letter 72. Signature: III/93/1724/247-48, Mappe 2. Brief von Werner Heisenberg von Max-Planck-Institut für Physik und Astrophysik an American Philosophical Society. Language: English.

Letter 73. Signature: III/93/1724/245-46, Mappe 2. Brief von Enrico Cantore an Werner Heisenberg. Language: English.

Letter 74. Signature: III/93/1724/243-44, Mappe 2. Brief von Enrico Cantore an Werner Heisenberg. Language: English.

Letter 75. Signature: III/93/1724/241-42, Mappe 2. Brief von Werner Heisenberg an Enrico Cantore. Language: German.

Letter 76. Signature: III/93/1724/237-38, Mappe 2. Brief von Enrico Cantore an Werner Heisenberg. Language: English.

Letter 77. Signature: III/93/1724/239-40, Mappe 2. Brief von Werner Heisenberg an Enrico Cantore. Language: German.

Letter 78. Signature: III/93/1014/1-176, Die Bedeutung des Schönen in der exakten Naturwissenschaft (Manuskripttitel) [The Meaning of Beauty in Exact Natural Science (1971) 75 pages]. Language: English (manuscript not included).

Letter 79. Signature: III/93/1724/235-36, Mappe 2. Brief von Enrico Cantore an Werner Heisenberg. Language: English.

Letter 80. Signature: III/93/1724/171-72, Mappe 1. Brief von Werner Heisenberg an Enrico Cantore. Language: German.

Letter 81. Signature: III/93/1724/173-234, Mappe 2. Brief von Enrico Cantore an Werner Heisenberg (paper attached). Language: English (attachment in English not included).

Letter 82. Signature: III/93/1724/169-70, Mappe 1. Brief von Werner Heisenberg an Enrico Cantore. Language: German.

Letter 83. Signature: III/93/1724/167-68, Mappe 1. Brief von Enrico Cantore an Werner Heisenberg. Language: English.

Letter 84. Signature: III/93/1724/165-66, Mappe 1. Brief von Werner Heisenberg an Enrico Cantore. Language: German.

Letter 85. Signature: III/93/1724/127-64, Mappe 1. Brief von Enrico Cantore an Werner Heisenberg (ref. III/93/1014/1-176). Language: English.

Letter 86. Signature: III/93/1724/125-26, Mappe 1. Brief von Werner Heisenberg an Enrico Cantore. Language: German.

Letter 87. Signature: III/93/1724/125-26, Mappe 1. Brief von Enrico Cantore an Werner Heisenberg. Language: English.

Letter 88. Signature: III/93/1724/121-22, Mappe 1. Brief von Enrico Cantore an Werner Heisenberg. Language: English. Handwritten.

Letter 89. Signature: III/93/1724/119-20, Mappe 1. Brief von Werner Heisenberg an Enrico Cantore. Language: German.

Letter 90. Signature: III/93/1724/117-18, Mappe 1. Brief von Enrico Cantore an Annemarie Giese. Language: German. Handwritten.

Letter 91. Signature: III/93/1724/113-16, Mappe 1. Brief von Enrico Cantore an Werner Heisenberg. Language: English.

Letter 92. Signature: III/93/1724/109–12, Mappe 1. Brief von Werner Heisenberg an Enrico Cantore. Language: German.
Letter 93. Signature: III/93/1724/101–8, Mappe 1. Brief von Enrico Cantore an Werner Heisenberg. Language: English.
Letter 94. Signature: III/93/1724/99–100, Mappe 1. On the book 'Man and Science' by Enrico Cantore (Manuskripttitel). Language: English.
Letter 95. Signature: III/93/1724/59–98, Mappe 1. Brief von Enrico Cantore an Werner Heisenberg. Language: English.
Letter 96. Signature: III/93/1724/39–58, Mappe 1. Brief von Enrico Cantore an Werner Heisenberg. Language: English.
Letter 97. Signature: III/93/1724/37–38, Mappe 1. Brief von Werner Heisenberg an Unbekannt o.D. Language: English.
Letter 98. Signature: III/93/1724/31–36, Mappe 1. Brief von James F. Mathias von John Simon Guggenheim Memorial Foundation an Werner Heisenberg. Language: English.
Letter 99. Signature: III/93/1724/37–38, Mappe 1. Brief von Werner Heisenberg an John Simon Guggenheim Memorial Foundation. Language: English.
Letter 100. Signature: III/93/1724/29–30, Mappe 1. Brief von Enrico Cantore von Lowenstein Center at Fordham University. Institute for Scientific Humanism an Werner Heisenberg. Language: English (two attachments in English included).
Letter 101. Signature: III/93/1724/9–28, Mappe 1. Brief von Enrico Cantore von Lowenstein Center at Fordham University. Institute for Scientific Humanism an Werner Heisenberg. Language: English (attachment in English included)
Letter 102. Signature: III/93/1724/7–8, Mappe 1. Brief von Enrico Cantore von Lowenstein Center at Fordham University. Institute for Scientific Humanism an Werner Heisenberg. Language: English.
Letter 103. Signature: III/93/1724/5–6, Mappe 1. Brief von Werner Heisenberg an Enrico Cantore an Lowenstein Center at Fordham University. Institute for Scientific Humanism. Language: English.
Letter 104. Signature: III/93/1724/1–2, Mappe 1. Brief von Enrico Cantore von Lowenstein Center at Fordham University. Institute for Scientific Humanism an Werner Heisenberg. Language: English.
Letter 105. Signature: III/93/1724/3–4, Mappe 1. Brief von Enrico Cantore von Lowenstein Center at Fordham University. Institute for Scientific Humanism an Werner Heisenberg. Language: English.
Letter 106. Signature: III/93/1661, Brief von Enrico Cantore von Lowenstein Center at Fordham University. Institute for Scientific Humanism an Annemarie Giese. Language: English.
Letter 107. Signature: III/93/1661, Brief von Enrico Cantore von Lowenstein Center at Fordham University. Institute for Scientific Humanism an Annemarie Giese. Language: English.

Books and Articles

Archie, Lee C. Review of *Scientific Man: The Humanistic Significance of Science* by Enrico Cantore. *American Scientist* 67 (1979) 121.

"Atomic Pioneer Offered Uncertainty Principle as a Basic Theory." *New York Times*, February 2, 1976. 25.

Binns, Peter. Review of *Atomic Order: An Introduction to the Philosophy of Microphysics* by Enrico Cantore. *British Journal for the Philosophy of Science* 22 (1971) 198–200.

Born, Max. *My Life and Views*. New York: Scribner, 1968.

Brendel, Bettina. "The Influence of Atomic Physics on My Paintings." *Leonardo* 6 (1973) 137–39.

Büchel, Wolfgang. *Philosophische Probleme der Physik*. Freiburg: Herder, 1965.

———. *Wille, Wunder, Welt: Physikalisches Weltbild und christlicher Glaube*. 2nd ed. Kevelaer, Germany: Butzon und Bercker, 1962.

Bunge, Mario. Review of *Der Teil und das Ganze* by Werner Heisenberg. *Physics Today* 24 (1971) 63–64.

Camilleri, Kristian. "Heisenberg and the Transformation of Kantian Philosophy." *International Studies in the Philosophy of Science* 19 (2005) 271–87.

Cantore, Enrico. "Atomic Order: An Introduction to the Philosophy of Microphysics." PhD diss., Pontifical Gregorian University, 1966.

———. *Atomic Order: An Introduction to the Philosophy of Microphysics*. Cambridge: MIT Press, 1969.

———. "The Christic Origination of Science." *Journal of the American Scientific Association* 37 (1985) 211–22.

———. "Genetical Understanding of Science: Some Considerations About Optics." *Archives Internationales d'Histoire des Sciences* 19 (1966) 333–63.

———. "Human Dignity, Science and Growth: The Humanistic Nature of Development." *WISH Basic Papers* 1 (1981) 1–24.

———. "Humanistic Significance of Science: Some Methodological Considerations." *Philosophy of Science* 38 (1971) 395–412.

———. "The Humanity of Science: A Philosophical Response to a Thought-Provoking Essay." *Atti della Fondazione Giorgio Ronchi* 28 (1973) 165–86.

———. "The Italian Philosophical Encyclopedia." *Review of Metaphysics* 24 (1971) 510–32.

———. "Leadership for Human Dignity: The Developmental Challenge to Scientific Professionals." In *Issues of Development: Towards a New Role for Science and Technology*, edited by Maurice Goldsmith and Alexander King, 247–57. Oxford: Pergamon, 1979.

———. "My Humanistic-Scientific Apostolate: A Jubilee Testimony." Unpublished paper, 1993.

———. "Per una integrazione umanizzante tra scienza e uomo." *Civiltà Cattolica* 125 (1974) 322–36.

———. "The Philosophical Problems Arising from Quantum Mechanics." Laurea thesis, University of Turin, 1955.

———. "Philosophy in Atomic Physics: Complementarity." *Modern Schoolman* 34 (1957) 79–104.

———. Review of *Physics and Beyond: Encounters and Conversations* by Werner Heisenberg. *Theological Studies* 32 (1971) 516–18.

———. "La sapienza biblica, ideale religioso del credente. I. Analisi del concetto ebraico di sapienza." *Rivista Biblica* 8 (1960) 1–9.

———. "La sapienza biblica, ideale religioso del credente. II. Aspetti intellettuali della sapienza biblica e loro evoluzione." *Rivista Biblica* 8 (1960) 129–43.

———. "La sapienza biblica, ideale religioso del credente. III. La sapienza biblica come conoscenza e timore-amore di Dio." *Rivista Biblica* 8 (1960) 193–205.

———. "Science and Humanism: The Sapiential Role of Philosophy." *Dialectica* 24 (1970) 215–41.

———. "Science as Dialogical Humanizing Process: Highlights of a Vocation." *Dialectica* 25 (1971) 293–316.

———. "Science for Human Dignity: A Christian Leadership Task." *Homiletic and Pastoral Review* 82 (1982) 47–53.

———. "Science, Religion and Human Dignity: Notes for Contemporary Evangelization." *Euntes Docete* 35 (1982) 213–24.

———. "Scientific Humanism and the University." *Thought* 43 (1968) 409–28.

———. *Scientific Man: The Humanistic Significance of Science*. New York: ISH, 1977.

———. "La scienza come fattore umanistico." Translated by G. Tassani and S. Marmi. *Il Regno-attualità* 10 (1982) 216–19.

———. "La scienza e l'uomo. Significato della crisi umanistica contemporanea." *Civiltà Cattolica* 125 (1974) 112–29.

———. "La soluzione fisica dell'enigma quantistico." *Divus Thomas* 60 (1957) 150–59.

———. "Some Reflections on Man's Unending Quest for Understanding." *Dialectica* 22 (1968) 132–66.

———. "Umanesimo Scientifico." In vol. 1 of *Dizionario Interdisciplinare di Scienza e Fede*, edited by Giuseppe Tanzella-Nitti and Alberto Strumia, 1399–409. Rome: Urbaniana University-Città Nuova, 2002.

———. "Un Rapporto sul mio Apostolato: Primavera 2006." Unpublished paper.

Cassidy, David C. *Uncertainty: The Life and Science of Werner Heisenberg*. New York: Freeman, 1992.

Clarke, William Norris. Review of *Scientific Man: The Humanistic Significance of Science* by Enrico Cantore. *Religious Studies* 15 (1979) 130–31.

Courant, Richard, and Herbert Ellis Robbins. *What Is Mathematics?* London: Oxford University Press, 1941.

D'Agostino, Salvatore. Review of *Atomic Order: An Introduction to the Philosophy of Microphysics* by Enrico Cantore. *Science* 171 (1971) 890.

Dialectica. Review of *Atomic Order: An Introduction to the Philosophy of Microphysics* by Enrico Cantore. *Dialectica* 24 (1970) 325–26.

Forman, Paul. "Historiographic Doubts." Edited by Werner Heisenberg and Arnold J. Pomerans. *Science* 172 (1971) 687–88.

———. Review of *Atomic Order: An Introduction to the Philosophy of Microphysics* by Enrico Cantore. *Isis* 61 (1970) 535–36.

Francis. "*Veritatis Gaudium*: Apostolic Constitution on Ecclesiastical Universities and Faculties." *Acta Apostolicae Sedis* 110 (2018) 137–59.

Guardini, Romano. *Letters from Lake Como: Explorations on Technology and the Human Race*. Grand Rapids: Eerdmans, 1994.

Heelan, Patrick A. "Phenomenology, Ontology, and Quantum Physics." *Foundations of Science* 18 (2011) 379–85.

———. *Quantum Mechanics and Objectivity: A Study of the Physical Philosophy of Werner Heisenberg*. Leiden: Nijhoff, 1965.

Heisenberg, Werner. *Across the Frontiers*. New York: Harper & Row, 1974.

———. "Die Bedeutung des Schönen in der exakten Naturwissenschaft." *Physikalische Blätter* 27 (1971) 97–107.

---. "Die Bedeutung des Schönen in der exakten Naturwissenschaft/The Meaning of Beauty in Exact Natural Science." Translated by Enrico Cantore. Lithographs by Max Ernst. Stuttgart: Belser-Presse, 1971.

---. "Development of Concepts in the History of Quantum Theory." *American Journal of Physics* 43 (1975) 389–94.

---. "Double Dialogue with Werner Heisenberg." 1974. In vol. 3 of *Philosophical and Popular Writings*, edited by W. Blum et al., 464–86. Gesammelte Werke—Collected Works Series C. Munich: Piper, 1985.

---. "Grundlegende Voraussetzungen in der Physik der Elementarteilchen." In *Martin Heidegger zum siebzigsten Geburtstag: Festschrift*, edited by Günther Neske, 291–97. Pfullingen: Neske, 1959.

---. *Natural Law and the Structure of Matter*. London: Rebel, 1970.

---. "Naturwissenschaftliche und religiöse Wahrheit." *Universitas* 16 (1974).

---. *Ordnung der Wirklichkeit*. Munich: Piper, 1942.

---. *Philosophic Problems of Nuclear Science: Eight Lectures*. Translated by F. C. Hayes. London: Faber and Faber, 1952.

---. *Physics and Beyond*. New York: Harper & Row, 1971.

---. *Physics and Philosophy: The Revolution in Modern Science*. New York: Harper & Row, 1958.

---. "Planck's Discovery and the Philosophical Problems of Modern Physics." In *On Modern Physics*, 9–28. New York: Collier, 1961.

---. "Religion and Science (1927)." In *Physics and Beyond*, 82–92. New York: Harper & Row, 1971.

---. *Schritte über Grenzen: Gesammelte Reden und Aufsätze*. Munich: Piper, 1971.

---. "Scientific Truth and Religious Truth." *CrossCurrents* 24 (1975) 463–73.

---. "The Teachings of Goethe and Newton on Colour in the Light of Modern Physics." In *Philosophic Problems of Nuclear Science: Eight Lectures*, 67–86. London: Faber and Faber, 1952.

---. *Der Teil und das Ganze: Gespräche im Umkreis der Atomphysik*. Munich: Piper, 1969.

---. *Wandlungen in den Grundlagen der Naturwissenschaft*. Leipzig: Hirzel, 1934.

---. "Wolfgang Pauli's Philosophical Views." Translated by Kurt Leidecker. *Main Currents in Modern Thought* 17 (1961) 51–54.

---. "Wolfgang Paulis Philosophische Auffassungen." *Die Naturwissenschaften* 46 (1959) 661–63.

Heisenberg, Werner, and Elisabeth Heisenberg. *My Dear Li: Correspondence, 1937–1946*. Edited by Anna Maria Hirsch-Heisenberg. Translated by Irene Heisenberg. New Haven, CT: Yale University Press, 2016.

Hufbauer, Karl. Review of *Physics and Beyond: Encounters and Conversations* by Werner Heisenberg. *Isis* 62 (1971) 558–60.

John Paul II. "Address to the Pontifical Academy of Sciences, November 13, 2000." In vol. 23.2 of *Insegnamenti di Giovanni Paolo II*, 874–78. Vatican City: Libreria Editrice Vaticana, 2001.

Kane, Robert H. Review of *Atomic Order: An Introduction to the Philosophy of Microphysics* by Enrico Cantore. *Review of Metaphysics* 23 (1970) 739.

Landsberg, Peter Theodore. Review of *Scientific Man: The Humanistic Significance of Science* by Enrico Cantore. *Leonardo* 12 (1979) 72.

Monod, Jacques. *Chance and Necessity: An Essay on the Natural Philosophy of Modern Biology.* Translated by Austryn Wainhouse. New York: Knopf, 1971.

———. *Le Hasard et la Nécessité: Essai sur la philosophie naturelle de la biologie moderne.* Paris: Éditions du Seuil, 1970.

Morison, Elting Elmore. "Poker and Mozart Helped with the Atom." *New York Times Book Review,* January 17, 1971. 1.

Morrison, Philip. Review of *Physics and Beyond: Encounters and Conversations* by Werner Heisenberg. *Scientific American* 224 (1971) 127–28.

Newman, John Henry. *The Idea of a University.* Notre Dame, IN: University of Notre Dame, 1981.

Rabi, Isidor Isaac. *My Life and Times as a Physicist.* Claremont, CA: Claremont College, 1960.

———. *Science: The Center of Culture.* New York: World, 1970.

Russo, François. Review of *Atomic Order: An Introduction to the Philosophy of Microphysics* by Enrico Cantore. *Archives de Philosophie* 37 (1974) 156–60.

———. Review of *Scientific Man: The Humanistic Significance of Science* by Enrico Cantore. *Archives de Philosophie* 41 (1978) 306–8.

Selvaggi, Filippo. Review of *Scientific Man: The Humanistic Significance of Science* by Enrico Cantore. *Gregorianum* 58 (1977) 778–80.

Shelley, Thomas J. *Fordham: A History of the Jesuit University of New York: 1841–2003.* New York: Fordham University Press, 2016.

Shinn, Roger L. Review of *Scientific Man: The Humanistic Significance of Science. Theological Studies* 39 (1978) 808–10.

Skinner, B. F. *Beyond Freedom and Dignity.* New York: Knopf, 1971.

———. *Walden Two.* Indianapolis, IN: Hackett, 1948.

Smith, Cyril S. *Sources for the History of the Science of Steel, 1532–1786.* Cambridge, MA: Society for the History of Technology, 1968.

Snow, Charles Percy. *The Two Cultures and a Second Look: An Expanded Version of The Two Cultures and the Scientific Revolution.* Cambridge: Cambridge University Press, 1964.

Thomas, Robert McGill, Jr. "Leo McLaughlin, Jesuit Teacher, Dies at 84." *New York Times,* August 18, 1996. 51.

Trinklein, Frederick. *The God of Science.* Grand Rapids: Eerdmans, 1971.

Vatican Council II. "*Gaudium et Spes*: Pastoral Constitution on the Church in the Modern World." In *The Documents of Vatican II,* edited by Walter M. Abbott, 199–308. New York: America, 1966.

Weizsäcker, Carl Friedrich von. "Heisenbergs Entwicklung seit 1927." In *Werner Heisenberg,* edited by Carl Friedrich von Weizsäcker and B. L. van der Waerden, 1–15. Munich: Hanser, 1977.

Wertheimer, Max. *Productive Thinking.* New York: Harper, 1945.

———. *Produktives Denken.* Translated by Wolfgang Metzger. Frankfurt: Kramer, 1957.

Whitbeck, Caroline. Review of *Atomic Order: An Introduction to the Philosophy of Microphysics* by Enrico Cantore. *American Scientist* 58 (1970) 677.

www.ingramcontent.com/pod-product-compliance
Lightning Source LLC
Chambersburg PA
CBHW021650230426
43668CB00008B/576